Guidance note on inspection and testing

2nd edition

Published by: The Institution of Electrical Engineers, Savoy Place, London, United Kingdom, WC2R 0BL

© 1992: The Institution of Electrical Engineers

Issued August 1992
Reprinted April 1993, with amendments (Section 17)
2nd edition incorporating Amendment No. 1 to BS 7671 —
June 1995

Copies may be obtained from:

The Institution of Electrical Engineers
PO Box 96, Stevenage
Herts SG1 2SD

Tel: 01438 767242
Fax: 01438 742792

ISBN 0 85296 867 1

Contents

Preface

This Guidance Note is part of a series issued by the Wiring Regulations Committee of the Institution of Electrical Engineers to enlarge upon and amplify some of the requirements of BS 7671: 1992 Requirements for electrical installations (IEE Wiring Regulations Sixteenth Edition); incorporating Amendment No. 1 1994. Significant changes made in this 2nd edition are sidelined.

The scope generally follows that of the Regulations and the principal section numbers are shown on the left. The relevant Regulations and Appendices are noted in the right hand margin. Some Guidance Notes also contain material not included in BS 7671 (IEE Wiring Regulations) but which was included in earlier editions. All of the Guidance Notes contain references to other relevant sources of information.

Electrical installations in the United Kingdom which comply with BS 7671 are likely to satisfy Statutory Regulations such as the Electricity at Work Regulations 1989, but this cannot be guaranteed. It is stressed that it is essential to establish which Statutory and other Regulations apply and to install accordingly. For example an installation in premises subject to licensing may have requirements different from, or additional to BS 7671, and the legislation will take precedence.

Introduction

This Guidance Note is concerned with Part 7 — Inspection and Testing.

Neither the Wiring Regulations nor the Guidance Notes are design guides. It is essential to prepare a full specification prior to commencement or alteration of an electrical installation. The specification should set out the detailed design and provide sufficient information to enable competent persons to carry out the installation and to commission it. The specification must include a description of how the system is to operate and all the design and operational parameters. It must provide for all the commissioning procedures that will be required and for the provision of adequate information to the user. This will be by means of an operational manual or schedule.

120-01-02

514-09

It must be noted that it is a matter of contract as to which person or organisation is responsible for the production of the parts of the design, specification and any operational information.

The persons or organisations who may be concerned in the preparation of the specification include:

The Designer
The Installer
The Supplier of Electricity
The Installation Owner and/or User
The Architect
The Fire Prevention Officer
Any Regulatory Authority
Any Licensing Authority
The Health and Safety Executive

In producing the specification advice should be sought from the installation owner and/or user as to the intended use. Often, as in a speculative building, the intended use is unknown. The specification and/or the operational manual must set out the basis of use for which the installation is suitable.

Precise details of each item of equipment should be obtained from the manufacturer and/or supplier and compliance with appropriate standards confirmed.

120-04
120-05
511

The operational manual must include a description of how the system as installed is to operate and all commissioning records. The manual should also include manufacturers' technical data for all items of switchgear, luminaires, accessories, etc and any special instructions that may be needed. The Health and Safety at Work etc Act 1974 Section 6 is concerned with the provision of information, and guidance on the preparation of technical manuals is given in BS 4884 (Specification for technical manuals) and BS 4940 (Recommendations for the presentation of technical information about products and services in the construction industry). The size and complexity of the installation will dictate the nature and extent of the manual.

Part 1 — Initial Verification
Section 1 — The General Requirements

1.1	**Scope** **71**	This Part makes recommendations for the initial verification (inspection and testing) of electrical installations.

1.2 Safety
130

BS 7671: 1992 Requirements for Electrical Installations (IEE Wiring Regulations) requires that every new installation is to be inspected and tested to verify, as far as it is reasonably practicable, that the requirements of the Regulations have been met. — 130-10-01

Electrical testing inherently involves some degree of hazard; it is the tester's duty to ensure his own safety, and that of others, in the performance of his test procedures. The safety procedures detailed in Guidance Note GS 38 revised 'Electrical test equipment for use by electricians' of the Health and Safety Executive should be observed.

When using test instruments, safety is best achieved by precautions such as:

(i) understanding the equipment to be used and its rating

(ii) checking that all safety procedures are followed

(iii) checking that the instruments being used conform to the appropriate British Standard safety specifications. These are BS EN 61010 Safety requirements for electrical equipment for measurement, control, and laboratory use and BS 5458: 1978 (1993) Specification for safety requirements for indicating and recording electrical measuring instruments and their accessories

(iv) checking that test leads including any prods or clips used are in good order, are clean and have no cracked or broken insulation. Where appropriate, the requirements of the Health and Safety Executive Guidance Note GS 38 should be observed for test leads. This recommends the use of fused test leads aimed primarily at reducing the risks associated with arcing under fault conditions.

Particular attention should be paid to the safety aspects associated with any tests performed with instruments capable of generating a test voltage greater than 50 V, or which use the supply voltage for the purposes of the test in earth loop testing and residual current device (rcd) testing. Note the warnings given in Sections 15 and 16 of this Guidance Note.

Electric shock hazards can arise from, for example, capacitive loads such as cables charged in the process of an insulation test, or voltages on the earthed metalwork whilst conducting a loop test or rcd test. The test limits quoted in these guidelines are intended to minimise the chances of receiving an electric shock during tests.

1.3	**Frequency of subsequent inspections 130 154**	An assessment is to be made regarding the period between the initial test and the first periodic inspection. This is to be noted on the Completion Certificate, and on the Notice provided at the origin of the installation.	130-10-01 514-12
		According to the site conditions and the relevant experience of the inspector, the period between the initial, the first and subsequent periodic inspections and tests may be varied from the recommendations in the list of typical installations given in Part 2 of this Guidance Note.	732-01-01

2

1.4	**Inspector's qualifications 712**	The inspector carrying out the inspection and testing of any electrical installation must be competent. He must therefore be skilled, experienced and have sufficient knowledge of the type of installation to be inspected and tested to ensure no danger occurs to any person, livestock or property.

711-01-01
712-01-02

1.5 Information for the Inspector 311 513 711

The Regulations require that the following information shall be made available to the person or persons carrying out the inspection and test:

711-01-02

(i) the maximum demand, expressed in amperes after diversity is taken into account

311-01

(ii) the number and type of live conductors of the source of energy and of the circuits used in the installation

312-02

312-03

(iii) the type of earthing arrangement used by the installation and any facilities provided by the supplier for the user.

The information below regarding the basis of the design must be made available. This is also required by the Health and Safety at Work etc Act:

514-09-01

(iv) the type and composition of circuits, including points of utilisation, number and size of conductors and type of cable. This should include the Reference Installation method shown in Appendix 4 (paragraph 8) of BS 7671

(v) a description and the location of devices performing the functions of protection and/or isolation

(vi) the method selected to prevent danger from shock in the event of an earth fault

(vii) any circuit or equipment vulnerable to a particular test should be identified.

It is a requirement of the Regulations that this information will be available as an on-site record of the design.

Section 2 — Initial Inspection

2.1	**General procedure 711**	Inspection, and where appropriate testing, should be carried out progressively throughout the different stages of erection and before the installation is put into service.

2.2	**Scope 712**	Chapter 71 of the Regulations states the requirements for 'INITIAL VERIFICATION'. As far as reasonably practicable, an inspection shall be carried out to verify:

(i) all fixed equipment and material is of the correct type and complies with applicable British Standards or acceptable equivalents

(ii) all parts of the fixed installation are correctly selected and erected

(iii) no part of the fixed installation is visibly damaged or otherwise defective

(iv) the suitability of the installation relative to the environmental conditions.

2.3	**Inspection 712-01-03**	The inspection shall include at least the checking of those items listed in Section 712 of BS 7671.

712-01-03

Special attention should be paid to the inspection checklist in paragraph 2.5 (page 13), which provides a useful guide during installation and inspection.

2.3.1
712-01-03(i)

Connection of conductors　　　526

Every connection between conductors and equipment/other conductors shall provide durable electrical continuity and adequate mechanical strength. Requirements for the means, enclosure and accessibility of connections must be considered.

2.3.2
712-01-03(ii)

Identification of each conductor or non-flexible cable

Table 51A of the Wiring Regulations provides a schedule of Colour Identification of each core of non-flexible cable and bare conductor.　　　514

It should be checked that each core or bare conductor is identified as necessary. Busbar and pole colour should also comply with Table 51A.

Where it is desired to indicate phase rotation, or a different function for cables of the same colour, numbered or lettered sleeves are permitted.

2.3.3
712-01-03(ii)

Identification of flexible cable and cord

Protective conductors only shall be identified by a combination of green and yellow colours. Note particularly the identification in Table 51B of the Wiring Regulations for flexible cable with up to five cores.　　　514

2.3.4
712-01-03(iii)

Routeing of cables

Cable routes shall be selected with regard to the cables' suitability for the environment — ambient temperature, heat, water, foreign bodies, corrosion, impact, vibration, flora, fauna, radiation, building use and structure.　　　522

2.3.5 **712-01-03(iv)**	*Current-Carrying Capacity*	
	Where practicable, the cable size should be assessed against the protective arrangement based upon information provided by the installation designer (where available).	523,4,5
	Reference should be made, as appropriate, to Appendix 4 of the Wiring Regulations (installation methods).	

| **2.3.6**
712-01-03(v) | *Verification of Polarity — Single-Pole Device in a TN or TT System* | 130-05 |
| | It must be checked that no single-pole switch or protective device is installed in a neutral conductor. | 530 |

2.3.7 **712-01-03(vi)** **712-01-03(xiv)**	*Accessories and Equipment*	553
	Correct connection (suitability, polarity etc) is to be checked.	
	Table 55A of the Wiring Regulations is a schedule of types of plug and socket-outlets available, the rating, and the associated British Standard.	
	Particular attention should be paid to the requirements for a cable coupler.	553-02-01
	Bayonet lampholders should comply with BS 5042 and be of temperature rating T2.	553-03

| **2.3.8**
712-01-03(vii) | *Selection and Erection to Minimise the spread of fire* | 527-02
527-03 |
| | A fire barrier and/or protection against thermal effects should be provided if necessary to meet the requirements of the Wiring Regulations. | |

The Regulations require that each sealing   527-04
arrangement be inspected to verify that it conforms
with the manufacturer's erection instructions. This
may be impossible without dismantling the system
and it is essential, therefore, that inspection should be
carried out at the appropriate stage of the work, and
that this is recorded at the time for incorporation in
the inspection and test documents.

2.3.9
712-01-03
(viii)(a)   *Protection against direct and indirect contact*

SELV is the common method of providing protection   411-02
against both direct and indirect contact. The many
requirements include:

(a) an isolated source eg a safety isolating   411-02
     transformer to BS 3535

(b) electrical separation from higher voltage
     systems

(c) no connection with earth by the secondary
     circuits

(d) SELV exposed-conductive-parts shall have no
     connection with earth, exposed-conductive-parts
     or protective conductors of other systems.

2.3.10
712-01-03
(viii)(b)   *Protection against Direct Contact*   410

Insulation
Although protection by insulation is the usual   412-02
method, there are other methods of protection   471-04
against direct contact.

Barriers
Where live parts are protected by barriers, or   412-03
enclosures, these should be checked for adequacy and   471-05
security.

Obstacles
Protection by obstacles provides protection only
against unintentional contact. If this method is used
the area shall be accessible only to skilled persons or
to instructed persons under supervision. This method
of protection should not be used in locations of
increased shock risk.

412-04
471-06

Out of reach
Placing out of reach protects also against direct
contact. Increased distances are necessary where long
or bulky conducting objects are likely to be handled
in the vicinity.

412-05
471-07

Bare live parts are permitted in an area accessible only
to skilled persons, and the dimensions of passageways
should be checked against the guidance in Appendix
3 of the Memorandum of Guidance on the Electricity
at Work Regulations issued by the Health and Safety
Executive.

PELV
The requirements for PELV are as for SELV except that
the secondary circuits are earthed at one point and
exposed-conductive-parts may have connections with
earth, exposed-conductive-parts or protective
conductors of other systems.

471-14-01
471-14-02

2.3.11
712-01-03
(viii)(c)

Protection against Indirect Contact

The 'Methods of Protection against Indirect Contact'
are classified in a number of sub-sections in the
Wiring Regulations, and are:

(i) earthed equipotential bonding and automatic
disconnection of supply

(ii) use of Class II equipment

(iii) non-conducting location

(iv) earth-free local equipotential bonding

(v) electrical separation.

Methods (ii) (iii) and (iv) are specialised protection systems. It is essential that inspection work of such systems be carried out by persons competent in the discipline and having adequate information on the design of the system.

(i)	earthed equipotential bonding and automatic disconnection of supply (EEBADOS)	413-02 471-08-01

The presence, correct sizing, labelling and connection of appropriate protective conductors must be confirmed as follows: 542

— earthing conductors 542-03
413-02-02

— main equipotential bonding conductors 547-02
543-03

— circuit protective conductors 413-02-15
547-03

— supplementary bonding conductors Part 6

The earthing system must be determined ie

— PME earth (TN-C-S system)

— TN-S earth

— earth electrode

The earth impedance must be appropriate for the protective device ie rcd or overcurrent device. 413-02-04
413-02-08

Specialised Systems

(ii) use of class II equipment

This method of protection has very specialised requirements including continuing supervision to ensure the requirements continue to be met throughout the life of the installation.

(iii) non-conducting location

Protection by this method (as in a television test area) should be applied only in a special situation under supervision of a suitably qualified electrical engineer.

(iv)   earth-free local equipotential bonding

This method is applied in a special situation which is earth-free, under effective supervision, and where specified by a suitably qualified electrical engineer. A warning notice, complying with Section 514 must be fixed in a prominent position adjacent to every point of access to the location concerned. This method is sometimes combined with 'Electrical Separation'.

514-13-02

Inspection should verify that no item is earthed within the area and that no earthed services or conductors traverse the area, including the floor and ceiling. Inspection should confirm the designer's intentions and whether or not they have been achieved.

(v)   electrical separation

This method may be used when several items of equipment are supplied from a common source, and it must be specified by a suitably qualified electrical engineer. It should be noted that the requirements differ if only a single item of equipment is being supplied.

413-06
471-12

2.3.12
712-01-03(ix)

Prevention of Mutual Detrimental Influence

33
515
528

The requirements are stated in Chapter 33: Compatibility, and in Section 515: Mutual Detrimental Influence. Another aspect which should be taken into consideration during inspection is Section 528: Proximity to other Services.

2.3.13
712-01-03(x)

Isolating and Switching Devices

All switch utilization categories must be appropriate for the nature of the load — see Table 2.3.13. Checking of utilization category may need to be carried out during construction, if the label is obscured during erection. Guidance Note No. 2 (Isolation and Switching) provides more comprehensive guidance.

TABLE 2.3.13
Utilization categories for alternating current installations

Utilization category		Typical applications
Frequent operation	Infrequent operation	
AC-20A	AC-20B	- Connecting and disconnecting under no-load conditions
AC-21A	AC-21B	- Switching of resistive loads including moderate overloads
AC-22A	AC-22B	- Switching of mixed resistive and inductive loads, including moderate overloads
AC-23A	AC-23B	- Switching of motor loads or other highly inductive loads

Where practicable an isolation exercise should be carried out to check that effective isolation can be achieved. This should include, where appropriate, locking-off and inspection or testing to verify that the circuit is dead and no other source of supply is present.

2.3.14
712-01-03(xi)

Presence of under-voltage protective devices

Suitable precautions shall be taken where a reduction in voltage, or loss and subsequent restoration of voltage, can cause damage.

45

2.3.15
712-01-03(xii)

Protective Devices

The rating or setting of each protective device should be compared with the design.

| **2.3.16** | *Labelling of protective devices, switches and terminals* | 514-08 |
| **712-01-03(xiii)** | | |

A protective device shall be arranged and identified so that the circuit protected may be easily recognised.

| **2.3.17** | *Selection of equipment and protective measures* | 512-06 |
| **712-01-03(xiv)** | *appropriate to external influences* | 515 |

| **2.3.18** | *Adequacy of access to switchgear and equipment* | 130-07 |
| **712-01-03(xv)** | | 513 |

Every piece of equipment which requires operation or attention by a person shall be so installed that adequate and safe means of access and working space are afforded.

2.3.19	*Presence of danger notices and other warning notices*	514-10
712-01-03(xvi)		514-11
		514-12
		514-13

| **2.3.20** | *Presence of diagrams, instructions and similar* | 514-09 |
| **712-01-03(xvii)** | *information* | |

| **2.3.21** | *Erection methods* | Part 5 |
| **712-01-03(xviii)** | | |

Chapter 52 regulates on selection and erection in detail. Fixings of switchgear cables, conduit, fittings must be adequate for the environment.

2.4

Typical forms for use when inspecting and testing are included in Section 17.

2.5

Inspection Checklist

Listed below are requirements to be checked when carrying out an installation inspection. The list is not exhaustive.

<table>
<tr><td>General</td><td>

1. Complies with the requirements listed under 2.2 (130-01-01)
2. Accessible for operation, inspection and maintenance (513-01-01)
3. Suitable for local atmosphere and ambient temperature (BS 5345)
4. Circuits to be separate (no borrowed neutrals) (314-01-04)
5. Circuits to be identified (neutral and protective conductors in same sequence as phase conductors) (514-01-02, 514-08-01)
6. Protective devices adequate for intended purpose (Ch. 53)
7. Disconnection times likely to be met by installed protective devices (Ch. 41)
8. More than one socket provided for convenience (553-01-07)
9. All circuits suitably identified (514-09)
10. Suitable main switch provided (Ch. 46)
11. Supplies to any safety services suitably installed eg Fire Alarms BS 5839
12. Environmental IP requirements accounted for (BS EN 60529)
13. Means of isolation suitably labelled (Regulation 514-08)
14. Provision for disconnecting the neutral (460-01-06)
15. Double-pole isolation for domestic single-phase installations (476-01-03)
16. RCDs provided where required (417-08, 471-16)
17. Discrimination between rcds considered (BS 4293, 314-01-02)
18. Main earthing terminal provided (542-04-01) readily accessible and identified (513-13-01)
19. Provision for disconnecting earthing conductor (542-04-02)
20. Correct cable glands and gland-plates used (BS 6121)
21. Cables used meet requirements of relevant BS (Appendix 4 of the Regulations, 521-01-01)
22. Conductors correctly identified (Sect. 514)
23. Earth tail pots installed where required on mineral insulated cables (130-02-05)
24. Non-conductive finishes on enclosures removed to ensure good electrical connection (526-01-01)
25. Adequately rated fuseboard (BS 5486, may require derating)

</td></tr>
</table>

26. Correct fuses or circuit-breakers installed (Sects.
    531, 533)
27. All connections secure (130-01)
28. Consideration paid to electromagnetic effects
    and electromechanical stresses (Ch. 52)
29. Overload protection provided where applicable
    (Sect. 473 and Sect. 533)
30. Segregation of circuits (Sect. 515 and Sect. 528)
31. Retest notices provided (514-12-01)
32. Sealing of the wiring system including fire
    barriers (527-02)

Switchgear

1. Suitable for the purpose intended (Ch. 53)
2. Meets requirements of BS 4752, 5486 and BS 5345
   where applicable or equivalent standards (511-01)
3. Securely fixed (130-01) and suitably labelled (514-01)
4. Non-conductive finishes on switchgear removed
   at protection conductor connections (526-01-01)
5. Suitable cable glands and gland plates used
   (BS 6121)
6. Correctly earthed (Ch. 54)
7. Conditions likely to be encountered taken account
   of ie suitable for the foreseen environment
   (Sect. 522)
8. Correct IP rating applied (BS EN 60529)
9. Suitable as means of isolation, where applicable
   (Sect. 461)
10. Not accessible to person normally using a bath or
    shower (601-08)
11. Need for isolation, mechanical maintenance,
    emergency and functional switching met (Sect. 537)
12. Fireman's switch provided where required
    (537-04-06)
13. Switchgear suitably coloured where necessary
    (537-04-04)
14. All connections secure (Sect. 526)
15. Cables correctly terminated and identified
    (Sect. 514 and 526)
16. No sharp edges on cable entries, screw heads etc
    which could cause damage to cables (522-08)

Wiring accessories

1.   General (applicable to each type of accessory)
2.   Complies with BS 5733 or other appropriate standard (Sect. 511)
3.   Box or other enclosure securely fixed (130-01)
4.   Metal box or other enclosure earthed (471-08-08 and Sect. 54)
5.   Edge of flush box level with wall surface (526-03)
6.   No sharp edges on cable entries, screw heads etc which could cause damage to cables (522-08)
7.   Non-sheathed cables, and cores of cable from which sheath has been removed, not exposed outside the enclosure (526-03)
8.   Conductors correctly identified (514-06: 514-07)
9.   Bare protective conductors sleeved green/yellow (514-03)
10. Terminals tight and containing all strands of the conductors (Sect. 526)
11. Cord grip correctly used or clips fitted to cables to prevent strain on the terminals (522-08-05)
12. Adequate current rating (130-02-02)
13. Suitable for the conditions likely to be encountered (Sect. 522)

Lighting controls

1.   Light switches comply with BS 3676 or BS 5518 (Sect. 511)
2.   Readily accessible (Sect. 512)
3.   Single-pole switches connected in phase conductors only (530-01-02)
4.   Correct colour coding of conductors (514-06-01)
5.   Earthing of exposed metalwork, eg metal switch plate (Ch. 54)
6.   Switch out of reach of a person using a bath or shower (601-08)
7.   Adequate current rating (130-02-02)
8.   Suitable for inductive circuits or de-rated — where necessary (512-02)
9.   Switch labelled to indicate purpose, where this is not obvious (514-01)
10. Track systems comply with BS 4533 (521-06)
11. Appropriate controls suitable for the luminaires (553-04-01)

Lighting points

1. Ceiling rose complies with BS 67 (553-04-01)
2. Not more than one flex unless designed for multiple pendants (553-04-03)
3. Flex support devices used (553-04-04)
4. Switch wires identified (514-06-01)
5. Holes in ceiling above rose made good to prevent spread of fire (527-02-01)
6. Not connected to a supply exceeding 250 V (553-04-02)
7. Suitable for the mass suspended (554-01-01)
8. Lampholders to BS 5042 or BS EN 60238

Socket-outlets

1. Complies with BS 196, BS 546, BS 1363, BS 4343, BS EN 60309-2 (553-01)
2. Mounting height above the floor or working surface suitable (553-01-06)
3. Correct polarity (130-05-01 and 713-09)
4. Not installed in a bathroom or shower room (601-10)
5. Not within 2.5 m of a shower cubicle in a room other than a bathroom (601-10-03)
6. Controlled by a switch, where the supply is direct current (537-05-05)
7. Protected where mounted on a floor (Sect. 52)
8. Not used to supply a water heater having uninsulated elements (554-05-03)
9. Circuit protective conductor connected directly to the earthing terminal of the socket-outlet, on a sheathed wiring installation (543-02-07)
10. Earthing tail from the earthed metal box, on a conduit installation to the earthing terminal of the socket-outlet (543-02-07)

Joint box

1. Joints accessible for inspection (526-04)
2. Joints protected against mechanical damage (526-03-01)
3. All conductors correctly connected (526-01)

Fused connection unit

1. Out of reach of a person using a bath or shower (Sect. 601)
2. Correct rating and fuse (533-01)

Cooker control unit

1. Sited to the side and low enough for accessibility and to prevent flexes trailing across radiant plates (130-07-01)

Conduits

General

1. Securely fixed and adequately protected against mechanical damage (522-08)
2. Inspection fittings accessible (522-08-02)
3. Number of cables for easy draw not exceeded (522-08-01)
4. Solid elbows and tees used only as permitted (522-08-01 and 522-08-03)
5. Ends of conduit reamed and bushed (522-08)
6. Adequate boxes (522-08)
7. Unused entries blanked off where necessary (412-03)
8. Not containing unsuitable non-electrical pipes or tubes (Sect. 511)
9. Provided with drainage holes (522-03)
10. Radius of bends such that cables are not damaged (522-08-03)

Rigid metal conduit

1. Complies with BS 31, or BS 4568 Parts 1 and 2 (Sect. 511)
2. Connected to the main earth terminal (413-02-06)
3. Segregated from fixed metalwork of other services (413-02 and 528-02)
4. Phase and neutral cables bunched in the same conduit (521-02)
5. Conduit in damp and corrosive situations (522-03 and 522-05)
6. Maximum span between buildings without intermediate support (522-08 and see Guidance Note No.1 and On-Site Guide)

Rigid non-metallic conduit

1. Complies with BS 4607 Parts 1 and 2 and BS 6099 (521-04)
2. Ambient and working temperatures within permitted limits (522-01 and 522-02)
3. Provision for expansion and contraction (522-08)

4.   Boxes and fixings suitable for mass of luminaire suspended at expected temperature (522-01)

Flexible metal conduit

1.   Complies with BS 731 (521-04)
2.   Not used as sole protective conductor (543-02-01)
3.   Adequately supported (522-08)

Trunking

General

1.   Complies with BS 4678 (521-05)
2.   Securely fixed and adequately protected against mechanical damage (522-08)
3.   Selected and erected so that no damage is caused by ingress of water (522-03)
4.   Proximity to non-electrical services (528-02)
5.   Barriers to prevent excessive temperature rise in vertical trunking (527-02)
6.   Holes surrounding trunking made good (577-02)
7.   Category 2 circuits partitioned from Category 1 circuits or insulated for the highest voltage present (528-01 and relevant BS)
8.   Category 3 circuits partitioned from Category 1 circuits or wired in mineral-insulated metal-sheathed cables (528-01 and relevant BS)
9.   Common outlets for Category 1, 2 and 3 circuits provided with screens, barriers or partitions (528-01 and relevant BS)

Metal trunking

1.   Phase and neutral cables bunched in the same metal trunking (521-02)
2.   Protected against damp or corrosion (522-03 and 522-05)
3.   Earthed (413-02)
4.   Joints mechanically sound, and of adequate continuity where trunking is used as a protective conductor (543-02)

Insulated Cables

Non-flexible cables

1.   Correct type (521-01)
2.   Correct current rating (523-01)

3. Protected against mechanical damage and abrasion (522-08)
4. Cables suitable for low ambient temperature (522-01)
5. Non-sheathed cables protected by enclosure in conduit, duct or trunking (521-07)
6. Exposed to direct sunlight, of a suitable type (522-11)
7. Not run in lift shaft unless part of the lift installation and of the permitted type (528-02-06 and BS 5655)
8. Correctly selected and installed for use underground (522-06-03)
9. Correctly selected and installed for use on exterior walls etc (Sect. 522)
10. Correctly selected and installed for use overhead (521-01-03 and 522-08)
11. Internal radii of bends in accordance with (relevant BS and 522-08)
12. Correctly supported (522-08-04)
13. Selected for ambient temperature (522-01)
14. Not exposed to water etc unless suitable for such exposure (522-03)
15. Metal sheaths and armour earthed (543-02)
16. Metal sheaths and armour segregated from, or bonded to, extraneous fixed metalwork of other services (528-02-05)
17. Identified at terminations (514-06)
18. Joints and connections electrically and mechanically sound and adequately insulated (526-01 and 526-02)
19. All wires securely contained in terminals etc without strain (Sect. 526)
20. Enclosure of terminals (Sect. 526)
21. Glands correctly selected and fitted (526-01 and BS 6121)
22. Joints and connections mechanically sound and accessible for inspection, except as permitted otherwise (526-04)
23. Earthed concentric wiring including cne cables to be used only as permitted (546-02) and Electricity Supply Regulations 1988 (as amended)

Flexible cables and cords

1. Correct type (521-01-01)
2. Correct current rating (Sect. 523)

3.   Protected where exposed to mechanical damage
     (522-06 and 522-08)
4.   Suitably sheathed where exposed to contact with
     water (522-03) and corrosive substances (522-05)
5.   Protected where used for final connections to fixed
     apparatus etc (521-07 and 526-03-03)
6.   Selected for resistance to damage by heat (522-02)
7.   Segregation of Category 1 and Category 2 circuits
     (BS 6701 and Sect. 528)
8.   Category 3 circuits not contained in common cable
     with Category 1 circuits (BS 6701 and Sect. 528)
9.   Cores correctly identified throughout (514-07)
10. Prohibited core colours not used (514-07-02)
11. Joints to be made using a cable coupler
     (526-02-01)
12. Not used as fixed wiring except as permitted
     (521-01-04)
13. Final connections to portable equipment a
     convenient length and connected as stated
     (553-01-07)
14. Final connections to other current using equipment
     (Sect. 554)
15. Mass supported by pendants not exceeding values
     stated (554-01)

Protective conductors

1.   Cables incorporating protective conductors comply
     with the relevant BS (Sect. 511)
2.   Joints in metal conduit, duct or trunking comply
     with Regulations (543-03)
3.   Flexible conduit to be supplemented by a protective
     conductor (543-02-01)
4.   Minimum cross-sectional area of copper conductors
     (543-01)
5.   Copper conductors, other than strip, of 6 sq mm or
     less protected by insulation (543-03)
6.   Circuit protective conductor at termination of
     sheathed cables insulated with sleeving (543-03-02)
7.   Bare circuit protective conductor protected against
     mechanical damage and corrosion
     (542-03 and 543-03-01)
8.   Insulation, sleeving and terminations identified by
     colour combination green-and-yellow (514-03-01)
9.   Joints sound (526-01)
10. Main and supplementary bonding conductors of
     correct size (Ch. 54)

11. Circuit protective conductors not less than 4 mm^2 if not protected against mechanical damage (547-03-01)

Enclosures

General

1. Suitable degree of protection (IP Code) appropriate to external influences (412-03 and Sect. 522)

Section 3 — Initial Testing

3.1 **Initial testing** For guidance on Initial Testing refer to Part 3,
713 Section 7 Reference Tests, for the Initial and Periodic
testing of electrical installations.

3.2 **Completion** The Regulations require that a Completion 741-01-01
certificate Certificate in the form set out in Appendix 6 of
741 BS 7671 together with a schedule of test results shall 742-02-02
742 be given to the person ordering the work.

Section 742 requires that:

(i) any defects or omissions revealed by the 742-02-01
Inspector shall be made good before a
Completion Certificate is issued

(ii) the Completion Certificate shall be 742-02-02
signed by a competent person or persons
stating that to the best of their knowledge
and belief the installation has been designed,
constructed, inspected and tested in accordance
with BS 7671, any permissible deviations being
listed.

3.3 **Standard** Typical forms for use when carrying out inspections
forms and tests are included in Section 17.

Part 2 — Periodic Inspection and Testing
Section 4 — The General Requirements

4.1 Scope
110

This Part makes recommendations for the Periodic Inspection and Testing of electrical installations. It does not apply to installations listed in Regulation 110-02 of the Wiring Regulations.

110-02

This Part may be suitable as the basis of a specification or contract but all limitations and exclusions should be made clear and agreed by the parties to any contract. Particular attention is drawn to the necessity of recording these points in the Report and to the sampling techniques described.

744-01-02

4.2 Requirement for periodic inspection and testing
731

Periodic Inspection and Testing is necessary because all electrical installations deteriorate due to a number of factors such as damage, wear, tear, corrosion, excessive electrical loading, ageing and environmental influences. Consequently:

731-01-02

(i) legislation requires that certain electrical installations are maintained in a safe condition and therefore must be periodically inspected and tested —Table 4A

(ii) licensing authorities, public bodies, insurance companies, mortgagors and others may require periodic inspection and testing of electrical installations —Table 4A

(iii) additionally, Periodic Inspection and Testing should be considered:

(a) to assess compliance with the current IEE Wiring Regulations

(b) on a change of ownership or tenancy of the premises

(c) on a change of use of the premises

(d) after alterations or additions to the original installation

(e) because of any significant change in the electrical loading of the installation

(f) where there is reason to believe that damage may have been caused to the installation.

Reference to legislation and other documents is made below and it is vital that these requirements are ascertained before undertaking a Periodic Inspection and Test.

4.3 Frequency of periodic inspection and testing
732

All installations should be regularly inspected and as necessary tested. In some cases there is a statutory requirement for inspections at specified intervals but, in others, the period is left to the discretion of the designer, installer or the inspector.

732-01-01

Recommended intervals may become mandatory if incorporated in Local Authority conditions of licence for various types of premises, or as the result of subsequent legislation.

The intervals between inspections should be in accordance with the recommendations in the latest Periodic Inspection and Testing Report relating to the installation or, in its absence, the Completion Certificate. Where those recommendations are appropriate to the present use and condition of the installation, the interval between inspections should not exceed the appropriate period indicated in Table 4A and should be noted on Reports or Test Schedules.

732-01-01

The intervals between inspections in Table 4A are maximum, and need to be reviewed regularly.

The effects of wear and tear, environment and ageing may be predictable and the future need for inspection reliably anticipated.

This is not so of damage. An installation at risk of damage, particularly in public areas where it may not be reported, will require frequent inspection, not necessarily supplemented by testing. Records of all inspections and subsequent repairs should be retained.

TABLE 4A
Frequency of inspection

Type of Installation	Maximum period between Inspections	Reference (see notes below)
General		
Domestic	10 years	2
Commercial	5 years	1, 2
Educational establishments	5 years	1 2
Hospitals	5 years	1, 2
Industrial	3 years	1, 2
Buildings open to the public		
Cinemas	1 year	2, 7 8
Church installations under 5 years old	2 years	2, 3
Church installations over 5 years old	1 year	2, 3
Leisure complexes	1 year	1, 2,7
Places of public entertainment	1 year	1, 2, 7
Restaurants and hotels	1 year	
Theatres	1 year	2, 7
External installations		
Agricultural and Horticultural	3 years	1, 2
Caravans	3 years	2
Caravan parks	1 year	1, 2, 7
Highway power supplies	6 years	
Special installations		
Emergency lighting	3 years	2, 4, 5
Fire alarms	1 year	2, 5, 6
Launderettes	1 year	1, 2, 7
Petrol filling stations	1 year	1, 2, 7
Short term installations	3 months	1, 2
Construction site installations	3 months	1, 2

Reference Key

1. Particular attention must be taken to comply with SI 1988
   No 1057. The Electricity Supply Regulations 1988 (as amended).
2. SI 1989 No 635. The Electricity at Work Regulations 1989
   (Regulation 4 & Memorandum).
3. Lighting and Wiring of Churches, 1981. Church Information
   Office.
4. BS 5266.
5. Other intervals are recommended for testing operation of
   batteries and generators.
6. BS 5839.
7. Local Authority Conditions of Licence.
8. SI 1955 No 1129 (Clause 27). The Cinematograph (Safety)
   Regulations.

4.4 **Inspector's**   The Inspector carrying out the inspection and any   742-02-02
     qualifications necessary testing of any electrical installation must be
     742 competent. He must therefore be skilled, experienced
and have sufficient knowledge of the type of
installation to be inspected and tested to ensure no
danger occurs to any person, livestock or property.

4.5 **Information for** It is essential that the inspector knows the extent of
     the Inspector the installation to be inspected and the depth, or
criteria regarding the limit, of the inspection. This
should be recorded.

Enquiries should be made to the person responsible   514-09
for the electrical installation with regard to the
provision of diagrams, design criteria, electricity
supply and earthing arrangements.

Diagrams, charts or tables should be available to
indicate the type and composition of circuits,
identification of protective devices for shock
protection, isolation and switching and a description   413-02-04
of the method used for protection against indirect
contact.

Section 5 — Periodic Inspection

5.1 General procedure

Where diagrams, charts or tables are not available, a degree of exploratory work may be necessary so that inspection and testing can be carried out safely and effectively.

712-01-03 (xvii) 514-09

Note should be made of any known changes in environmental conditions, building structure, and alterations or additions which have affected the suitability of the wiring for its present load and method of installation.

A careful check should be made of the type of equipment on site so that the necessary precautions can be taken, where conditions permit, to disconnect or short-out electronic and other equipment which may be damaged by subsequent testing. Special care must be taken where control and protective devices contain electronic components.

713-04-02 713-04-06

It is essential to determine the degree of disconnection which will be acceptable before planning the detailed inspection and testing.

If the inspection and testing cannot be carried out safely without diagrams or equivalent information, they must be prepared if there is to be compliance with Section 6 of the Health and Safety at Work etc Act 1974.

514-09

5.2 Scope 731

With Periodic Inspection and Testing, **INSPECTION** is the **VITAL** initial operation and the testing subsequently carried out is in support of the Inspection.

731-01-02

For safety, it is necessary to carry out a visual inspection of the installation before beginning any tests or opening enclosures, removing covers, etc. So far as is reasonably practicable, the visual inspection must verify that the safety of persons, livestock and property is not endangered.

A thorough visual inspection should be made of all electrical equipment which is not concealed, and should include the accessible internal condition of a sample of the equipment. The external condition should be noted and if damage is identified or if the degree of protection has been impaired, the matter should be recorded on the schedule to the Report. This inspection should be carried out without power supplied to the installation, wherever possible, in accordance with the Electricity at Work Regulations 1989. The inspection should include a check on the condition of all electrical equipment and material, taking into account any available manufacturer's information, with regard to the following:

  (i) safety

  (ii) wear and tear

  (iii) corrosion

  (iv) damage

  (v) excessive loading (overloading)

  (vi) age

  (vii) external influences

(viii) suitability.

The assessment of condition should take account of known changes in conditions influencing and affecting electrical safety, eg extraneous-conductive-parts, plumbing, structural changes, etc.

5.3 Detailed inspection
7.3.1

The inspection should also include any special equipment where more detailed technical guidance may be necessary to identify factors influencing safety.

5.3.1   *Anticipation of danger*

During the inspection, the opportunity should be taken to identify dangers which might arise during the testing. Any location and equipment for which safety precautions may be necessary should be noted and the appropriate steps taken.

711
732-01-02

Enquiries should be made to establish the type of equipment on site so that precautions can be taken where necessary to disconnect electronic and other equipment which may be damaged by subsequent testing. Special care has to be taken where electronic components are fitted to protective and control devices. If inspection and testing cannot be carried out in any area, or is limited in any way, this should be recorded in the Report.

713-04-04

5.3.2   *Joints and Connections*

It is not practicable to inspect every joint and termination in an electrical installation. Nevertheless a sample inspection should be made. An inspection should be made of all accessible parts of the electrical installation eg switchgear, distribution boards, and a sample of luminaire points and socket-outlets to ensure that all terminal connections of the conductors are properly installed and secured. Any signs of overheating of conductors, terminations or equipment should be thoroughly investigated and included in the Report.

712-01-03
(i)
712-01-03
(vi)

5.3.3   *Conductors*

The means of identification of each conductor, including protective conductors, should be verified.

712-01-03
(ii)

The deterioration of, or damage to, conductors and their insulation, and their protective coverings, if any, should be noted.

The insulation and protective covering of each
conductor at each distribution board of the electrical
installation and at a sample of switchgear, luminaires,
socket-outlets and other points, should be inspected
to determine their condition and correct installation.
There should be no signs of overheating, overloading
or damage to the insulation, armour, sheath or
conductors.

712-01-03
(iv)

5.3.4   *Flexible Cables and Cords*

731   Where a flexible cable or cord forms part of the fixed
wiring installation, the inspection should include:

712-01-03
(iii)
514

   (i)   examination of the cable or cord for damage
or defects

  (ii)   examination of the terminations and
anchorages for damage or defects

(iii)   the correctness of its installation with regard
to additional mechanical protection, heat
resistant sleeving, etc.

5.3.5   *Switching Devices*

731   It is recommended that a random sample of a
minimum of 10 per cent of all switching devices is
given a thorough internal visual inspection of
accessible parts to assess their electrical and
mechanical conditions. Where the inspection reveals:

712-01-03
(v) (vi)

   (i)   results significantly different from results
recorded previously

  (ii)   results significantly different from results
reasonably to be expected

732   (iii)   adverse conditions, eg fluid ingress or worn
or damaged mechanisms.

The inspection should be extended to include every switching device associated with the installation under inspection unless there is clear evidence of how the damage occurred.

5.3.6   *Protection against thermal effects*

The presence of fire barriers and means of protection against thermal effects should be verified, if reasonably practicable.   712-01-03 (vii)

5.3.7   *Protection against direct and indirect contact*   712-01-03 (viii)(a)

SELV is commonly used as protection against direct and indirect contact. The requirements of 411-02 need to be checked, particularly with respect to the source eg a safety isolating transformer to BS 3535, the need to separate the circuits, and the segregation of exposed-conductive-parts of the SELV system from any connection with the earthing of the primary circuits, or from any other connection with earth.

411

411-02-02

411-02-06

411-02-07

5.3.8   *Protection against direct contact*

731   It should be established that the means of protection against direct contact with live conductors is provided by one or more of the following methods: 713-01-03 (viii)(b)
412

   (i)   insulation of live parts   713-04

   (ii) installation of barriers or enclosures   713-07

   (iii) obstacles

   (iv) placing out of reach

   (v) PELV.

It should be established that the means of protection against direct contact with any live conductors meets the requirements for the safety of any person, livestock and property from the effects of electric shock, fire and burns.

For each method of protection against direct contact it should be established that there has been no deterioration of insulation, no removal of barriers or obstacles and no alterations to enclosures or access to live conductors which would affect its effectiveness.

IT SHOULD BE NOTED THAT AN RCD MUST NOT BE USED AS A SOLE MEANS OF PROTECTION AGAINST DIRECT CONTACT WITH LIVE PARTS.

5.3.9 *Protection against indirect contact*

The method of protection against indirect contact must be determined and recorded.

712-01-03 (viii)(c) 413

For earthed equipotential bonding and automatic disconnection of supply the adequacy of main equipotential bonding and the connection of all protective conductors with the earth is essential. The loss of the earth connections will convert indirect contact to direct in the event of a fault.

5.3.10 *Protective devices*

712 The presence, accessibility, labelling and condition of devices for electrical protection, isolation and switching should be verified.

712-01-03 (x) 130-07

It should be established that each circuit is adequately protected with the correct type, size and rating of fuse or circuit-breaker. The suitability of each protective and monitoring device and its overload rating or setting should be checked.

712-01-03 (xii)

It must be ascertained that each protective device is correctly located and appropriate to the type of earthing system and to the circuits protected.

Each device for protection, isolation and switching should be readily accessible for normal operation, maintenance and inspection, and be suitably labelled where necessary.

712-01-03 (xv) 712-01-03 (xiii)

It should be established that a means of emergency
switching or where appropriate emergency stopping,
is provided where required by the Regulations, so
that the supply can be cut off rapidly to prevent or
remove danger. Where a risk of electric shock is
involved then the emergency switching device shall
switch all the live conductors except as permitted in
the Regulations.

713-01-03
(x)

When conditions permit, an internal inspection
should be made of any emergency switching device
and tests should be carried out as described later in
this Guide.

5.3.11   *Enclosures and mechanical protection*

713

The enclosure and mechanical protection of all
electrical apparatus and equipment should be
inspected to ensure that their condition remains
adequate for the type of protection intended.

713-07
130-08
712-01-03
(xiv)

5.3.12   *Marking and Labelling*

514
608

The labelling of each circuit should be verified.

514-02-01

It should be established that adjacent to every fuse or
circuit-breaker there is a label correctly indicating the
size and type of the fuse, nominal current of the
circuit-breaker and identification of the protected
circuit.

It should be ascertained that all switching devices are
correctly labelled and identify the circuits controlled.

Notices or labels are required at the following points
and equipment within an installation:

(i) at the origin of every installation

a notice of such durable material as to be likely
to remain easily legible throughout the life of the
installation, marked in indelible characters no
smaller than the example in the Regulations

IMPORTANT

This installation should be periodically inspected and tested and a report on its condition obtained, as prescribed in BS 7671 (formerly the IEE Wiring Regulations for Electrical Installations) published by the Institution of Electrical Engineers.

Date of last inspection.....................................

Recommended date of next inspection...................

shall be fixed in a prominent position at or near the origin of the installation:

(ii) where different voltages are present   514-10

    (a) equipment or enclosures within which a voltage exceeding 250 V exists and where the presence of such a voltage would not normally be expected

    (b) where terminals or other fixed live parts between which a voltage exceeding 250 V exists are housed in separate enclosures, or items of equipment which, although separated, can be reached simultaneously by a person

    (c) means of access to all live parts of switchgear and other fixed live parts where different nominal voltages exist.

(iii) earthing and bonding connections

712

    (a) a permanent label to BS 951 with the words   514-13-01

Safety Electrical Connection — Do Not Remove

shall be permanently fixed in a visible position at or near:

(1) the point of connection of every earthing conductor to an earth electrode

and

(2) the point of connection of every bonding
conductor to an extraneous-conductive-part

and

(3) the main earth terminal, where separate from
main switchgear.

(b) In an ***earth-free*** situation all exposed   514-13-02
metalwork should be bonded together but
not to the main earthing system. In this situation
a suitable permanent notice durably marked in
legible type and no smaller than the example in
the Wiring Regulations shall be permanently
fixed in a visible position.

> The equipotential protective bonding conductors
> associated with the electrical installation in this
> location **MUST NOT BE CONNECTED TO EARTH**.
> Equipment having exposed-conductive-parts
> connected to earth must not be brought into this
> location.

(iv) Residual Current Devices (rcds)

712

(a) Where rcds are fitted within an installation   514-12-02
a suitable permanent notice durably marked
in legible type and no smaller than the example
in the Wiring Regulations

> This installation, or part of it, is protected by a
> device which automatically switches off the
> supply if an earth fault develops. Test the
> device quarterly by pressing the button marked
> 'T' or 'Test'. If the device does not switch off
> the supply when the test button is pressed,
> seek expert advice.

shall be permanently fixed in a prominent position
at or near the main distribution board.

608   (v) Caravan installations

All touring (mobile) caravans shall have a
notice fixed near the main switch giving
instructions on the connection and
disconnection of the caravan installation to
the electricity supply.   608-07-05

The notice shall be of a durable material bearing
in indelible and easily legible characters and no
smaller than the text shown in the Wiring
Regulations, the text shown opposite.

5.3.13   *External influences*

Note should be made of any known changes in   712-01-01
external influences, building structure, and alterations   712-01-02
or additions which may have affected the suitability
of the wiring for its present load and method of
installation.

Note should be taken of any alterations or additions
of an irregular nature to the installation. If unsuitable
material has been used the Report should indicate
this together with reference to any evident faulty
workmanship or design.

5.3.14   *Enclosures and mechanical protection*

713   The enclosure and protection against mechanical   713-07
damage of all electrical apparatus and equipment   130-08
should be inspected to ensure that their condition
remains adequate for the type of protection intended.

INSTRUCTIONS FOR ELECTRICITY SUPPLY

TO CONNECT

1. Before connecting the caravan installation to the mains supply, check that:

    (a) the supply available at the caravan pitch supply point is suitable for the caravan electrical installation and appliances, and

    (b) the caravan main switch is in the OFF position.

2. Open the cover to the appliance inlet provided at the caravan supply point and insert the connector of the supply flexible cable.

3. Raise the cover of the electricity outlet provided on the pitch supply point and insert the plug of the supply cable.

THE CARAVAN SUPPLY FLEXIBLE CABLE MUST BE FULLY UNCOILED TO AVOID DAMAGE BY OVERHEATING

4. Switch on at the caravan main switch.

5. Check the operation of residual current devices, if any, fitted in the caravan by depressing the test buttons.

IN CASE OF DOUBT, OR IF AFTER CARRYING OUT THE ABOVE PROCEDURE THE SUPPLY DOES NOT BECOME AVAILABLE, OR IF THE SUPPLY FAILS, CONSULT THE CARAVAN PARK OPERATOR OR THE OPERATOR'S AGENT OR A QUALIFIED ELECTRICIAN

TO DISCONNECT

6. Switch off at the caravan main isolating switch, unplug both ends of the cable.

PERIODIC INSPECTION

Preferably not less than once every three years and more frequently if the vehicle is used more than normal average mileage for such vehicles, the caravan electrical installation and supply cable should be inspected and tested and a report on their condition obtained as prescribed in BS 7671 (formerly the Regulations for Electrical Installations) published by the Institution of Electrical Engineers.

Section 6 — Periodic Testing

6.1 Periodic testing general 731

Periodic tests should be made in such a way as to minimise disturbance of the installation and inconvenience to the user. Where it is necessary to disconnect part or the whole of an installation in order to carry out a test, the disconnection should be made at a time agreed with the user and for the minimum period needed to carry out the test. Where more than one test necessitates a disconnection where possible they should be made during one disconnection period.

Periodic testing may involve a greater degree of danger than initial testing and experienced professional advice should be sought.

Wherever a sample test indicates results significantly different from those previously recorded investigation is necessary. Unless the reason for the difference can be clearly identified as relating only to the sample tested, the size of the sample should be increased. If, in the increased sample, further failures to comply with the requirements of the Regulations are indicated, a 100 per cent test should be made.

The interval between tests should be confirmed at each test, see Table 4A.

6.2 Sequence of tests

Periodic testing of an installation includes the tests listed in Items (a) to (g) which shall be made in the sequence given. Reference methods of testing are provided in Part 3 of this Guidance Note but alternative methods which give no less effective results may be used.

    (a) continuity of all protective conductors (including earthed equipotential bonding conductors)

    (b) polarity

(c) earth fault loop impedance

(d) insulation resistance

(e) operation of devices for isolation and switching

(f) operation of residual current devices

(g) prospective fault current.

Where appropriate, the following tests must also be made:

(h) continuity of ring circuit conductors

(i) earth electrode resistance

(j) manual operation of overcurrent protective devices other than fuses

(k) electrical separation of circuits

(l) protection by non-conducting floors and walls.

These items are further discussed in Section 6.3.

6.3 Detailed periodic testing 713

6.3.1

Continuity of protective conductors and equipotential bonding conductors

If an electrical installation is isolated from the supply it is permissible to disconnect the protective and equipotential conductors from the main earthing terminal in order to verify their continuity. The sequence of operation needed for initial testing can be carried out safely on an existing installation if it is isolated from the supply.   713-02

Where an electrical installation cannot be isolated from the supply the protective and equipotential conductors must **NOT** be disconnected as, under fault conditions, the exposed and extraneous-conductive-parts could be raised to a dangerous level above earth potential.

The 'combined' integrity of the conductors shall be established by continuity/earth fault loop impedance tests. The former to establish continuity of conductors and the latter to establish and confirm an appropriate disconnection time in the event of an earth fault.

A continuity test should be made between all
exposed-conductive-parts likely to be touched in
normal use to ensure they are effectively earthed.

When testing the effectiveness of main bonding, the
resistance value between any service pipe or
extraneous-conductive-part and the main earthing
terminal should be of the order of 0.05 Ω or less.

Supplementary bonds should similarly have a
resistance of 0.05 Ω or less.

6.3.2　　　*Polarity*

713　　　Tests shall be made to verify that:

(i) the polarity is correct at the meter and
consumer unit/distribution board

(ii) single-pole control and protective devices are
connected in phase conductors only

(iii) conductors are correctly connected to socket-
outlets and other accessories/equipment

(iv) centre-contact bayonet and Edison-type
screw lamp-holders have their outer or
screwed contacts connected to the earthed
neutral conductor

(v) all multi-pole devices are correctly installed.

It should be established whether there have been any
alterations or additions to the installation since its last
inspection and test. If there have been no alterations
or additions then sample tests should be made of at
least 10 per cent of all single-pole and multi-pole
control devices and of any centre-contact
lampholders, together with 100 per cent of
socket-outlets. If any incorrect polarity is found then a
full test should be made in that part of the
installation supplied by the particular consumer unit
or distribution board concerned, and the sample
testing increased to 25 per cent for the remainder of
the installation. If additional cases of incorrect
polarity are found in the 25 per cent sample a full test
of the complete installation should be made.

731-01-02

713-09

721

6.3.3   *Earth fault loop impedance*

713   Where protective measures are used which require a knowledge of earth fault loop impedance, the relevant impedance shall be measured, or determined by an equally effective method.

Earth fault loop impedance tests are carried out at the locations indicated below:

(i)   at the origin of each installation and at each Distribution Board

(ii)   all fixed equipment and socket-outlets

(iii)   10 per cent (on a random basis) of all luminaires, with a minimum of one luminaire, preferably the furthest one from the consumer unit, for each circuit of any installation

(iv)   at any location which may be exposed to exceptional damage or deterioration or represent a special hazard

(v)   at the furthest point of every radial circuit.

Where the installation incorporates an rcd, the value of earth fault loop impedance obtained in the test should be related to the nominal residual operating current of the protective device, to verify compliance with Section 413.   413-02-17

6.3.4   *Insulation resistance*

713   Insulation resistance tests should be made on electrically isolated circuits and any electronic equipment which might be damaged by application of the test voltage must be isolated or only a measurement to protective earth made, with the phase and neutral connected together. Where practicable the tests are applied to the whole of the installation with all fuse links in place and all switches closed. Alternatively, the installation may be tested in parts, a distribution board at a time.   713-04

A dc voltage not less than that stated in Table 9.1 should be employed.

Table 9.1 of this Guide and 71A of the regulations require a minimum insulation resistance of 0.5 MΩ; however if the resistance is less than 2 MΩ further investigation is required to determine the cause of the low reading.

Where equipment is disconnected for these tests and equipment has exposed-conductive-parts required by the Regulations to be connected to protective conductors, the insulation resistance between the exposed-conductive-parts and all live parts of the equipment shall be measured separately and shall comply with the requirements of the appropriate British Standard for the equipment. If there is no appropriate British Standard for the equipment the insulation resistance shall be not less than 0.5 megohm.

6.3.5

Operation of devices for isolation and switching

712

Where means are provided in accordance with the requirements of the Regulations for isolation and switching, they shall be operated to verify their effectiveness and checked to ensure adequate and correct labelling.

712-12-02

514-01-01

For isolating devices in which the position of the contacts or other means of isolation is externally visible, visual inspection of operation is sufficient and no test need be made.

The operation of every safety switching device shall be checked by operating the device in the manner normally intended and determining that it performs its function correctly in accordance with the requirements of these Regulations.

Where it is a requirement that the device interrupts each of the supply conductors the test may necessitate the use of a test lamp or instrument connected between each phase and a connection to neutral on the load side of the switching device. Reliance should not be placed on a simple observation that the equipment controlled has ceased to operate.

Where switching devices are provided with detachable or lockable handles in accordance with the Regulations, a check should be made to verify that the handles or keys are not interchangeable with others used for similar purposes within the installation.

Where any form of interlocking is provided, eg between a mains circuit-breaker and an outgoing switch or isolation device, the integrity of the interlocking shall be verified.

Where switching devices are provided for isolation or for mechanical maintenance switching, the integrity of the means provided to prevent any equipment from being unintentionally or inadvertently re-energised shall be verified.

6.3.6 *Operation of Residual Current Devices to BS 4293*

Where protection is provided by an rcd or rcds, the effective operation of each rcd shall be verified by a test simulating an appropriate fault condition independent of any test facility incorporated in the device, followed by operation of the integral test device. The nominal rated tripping current for protection of a socket-outlet for use with equipment outdoors or installed outside the general installation equipotential zone must not exceed 30 mA. 713-12

Refer to Section 15 for detailed test procedure for General Purpose (non-delayed) rcds to BS 4293.

6.3.7 *Operation of Overcurrent Circuit-Breakers*

712 Where protection against overcurrent is provided by circuit-breakers the manual operating mechanism of each circuit-breaker should be operated to verify that the device opens and closes satisfactorily.

It is not normally necessary or practicable to test the
operation of the automatic tripping mechanism of
moulded case and miniature circuit-breakers. Any
such test would need to be made at a current
substantially exceeding the minimum tripping current
in order to achieve operation within a reasonable
time. For miniature circuit-breakers to BS 3871 a test
current not less than two and a half times the
nominal rated tripping current of the device is
needed for operation within 1 minute, and much
larger test currents are necessary to verify operation
of the mechanism for instantaneous tripping.

For miniature circuit-breakers and other circuit-
breakers of the sealed type, designed not to be
maintained, if there is doubt about the integrity of
the automatic mechanism it will normally be more
convenient to replace the device than to make further
tests. Such doubt may arise where, from the results of
visual inspection, the device appears to have suffered
damage or undue deterioration, or where there is
evidence that the device may have failed to operate
satisfactorily in service in the installation.

712-01-03

Circuit-breakers with the facility for injection testing
should be so tested and if appropriate relay settings
confirmed.

712-01-03
(x)

6.4 Periodic inspection and testing report
732
744

The Wiring Regulations require that the results of any
periodic inspection and test shall be recorded. A
written report shall be forwarded to the person
requesting the inspection.

732-01-03

Regulation 744 is more specific and requires a report
which contains the results and extent of a periodic
inspection, including:

744-01-02

(i) a description of the extent of the work, including
the parts of the installation inspected and details
of what the inspection and testing covered

(ii) details of any prevalent dangerous conditions or
non-compliance with the Regulations and any
which are likely to develop

(iii) any restrictions which may have been imposed during the inspection and testing of the installation

(iv) any serious defect, which should be remedied if possible. If not remedied, the defect should be reported to the Duty Holder (as defined in the Electricity at Work Regulations 1989)

(v) schedules of test results.

Installations constructed in accordance with earlier editions of the Wiring Regulations should be inspected and tested for compliance with the current edition and departures recorded. However reference should be made to the note by the Health and Safety Executive following the preface to BS 7671, that installations conforming to earlier editions and not complying with the current edition do not necessarily fail to achieve conformity with the Electricity at Work Regulations 1989.

731-01-02

The purpose of Periodic Inspection and Test is to ascertain whether the installation meets the general object of the Regulations, that is to say that persons, property and livestock are protected against electric shock, fire, burns and unexpected movement of machinery.

Part 3 — Reference Tests
Section 7 — The Sequence of Tests

7.1 Sequence of tests 713

Initial tests should be carried out in the following sequence: — 713-01-01

(i) Before the supply is connected, or with supply disconnected as appropriate

    (a) continuity of protective conductors, main and supplementary bonding

    (b) continuity of ring final circuit conductors

    (c) insulation resistance

    (d) site applied insulation

    (e) protection by separation of circuits

    (f) protection by barriers or enclosures provided during erection

    (g) insulation of non-conducting floors and walls

    (h) polarity

    (i) earth electrode resistance.

(ii) With the electrical supply connected re-check polarity before further testing

    (j) prospective fault current

    (k) earth fault loop impedance

    (l) residual current operated devices.

Periodic tests should be carried out in the sequence given in Section 6.2.

The test results should be recorded on an installation   130-10-01
schedule similar to form WR5 of Section 17. Forms of   732-01-03
completion or periodic inspection, inspection, test and
an installation schedule (including test results) should
be provided to the person ordering the work.

Section 8— Continuity of Protective Conductors

8.1 **Continuity of protective conductors main and supplementary bonding**

Every protective conductor including each bonding conductor shall be tested to verify that it is electrically sound and correctly connected.

The test methods detailed below, as well as checking the continuity of the protective conductor, also measure (R_1+R_2) which, when corrected for temperature, enables the designer to verify the calculated earth-fault loop impedance Z_S.

Instrument — Use a low-resistance ohmmeter for these tests. Refer to Section 16. 713-02-01

Test method 1

Connect the phase conductor to the protective conductor at the distribution board or consumer unit so as to include all the circuit. Then test between phase and earth terminals at each outlet in the circuit. The measurement at the circuit's extremity should be recorded and is the value of $(R_1 + R_2)$ for the circuit under test. 413-02-07 413-02-08

See Fig. 1 for Test method connections.

When testing the continuity of ring circuit protective conductors, the test should be carried out before connecting exposed-conductive-parts to the protective conductors, which may provide parallel paths. 713-03-01

Test method 2

Connect one terminal of the continuity tester to the consumer's earth terminal, and with a test lead from the other terminal make contact with the protective conductors at various points on the circuit, such as light fittings, switches, spur outlets etc.

The resistance reading obtained by the above method includes the resistance of the test leads. The resistance of the test leads should be measured and deducted from all resistance readings obtained using this method. The resistance of the protective conductor R_2 is recorded on the installation schedule, Form WR4.

To test the bonding conductors' continuity use Test method 2.

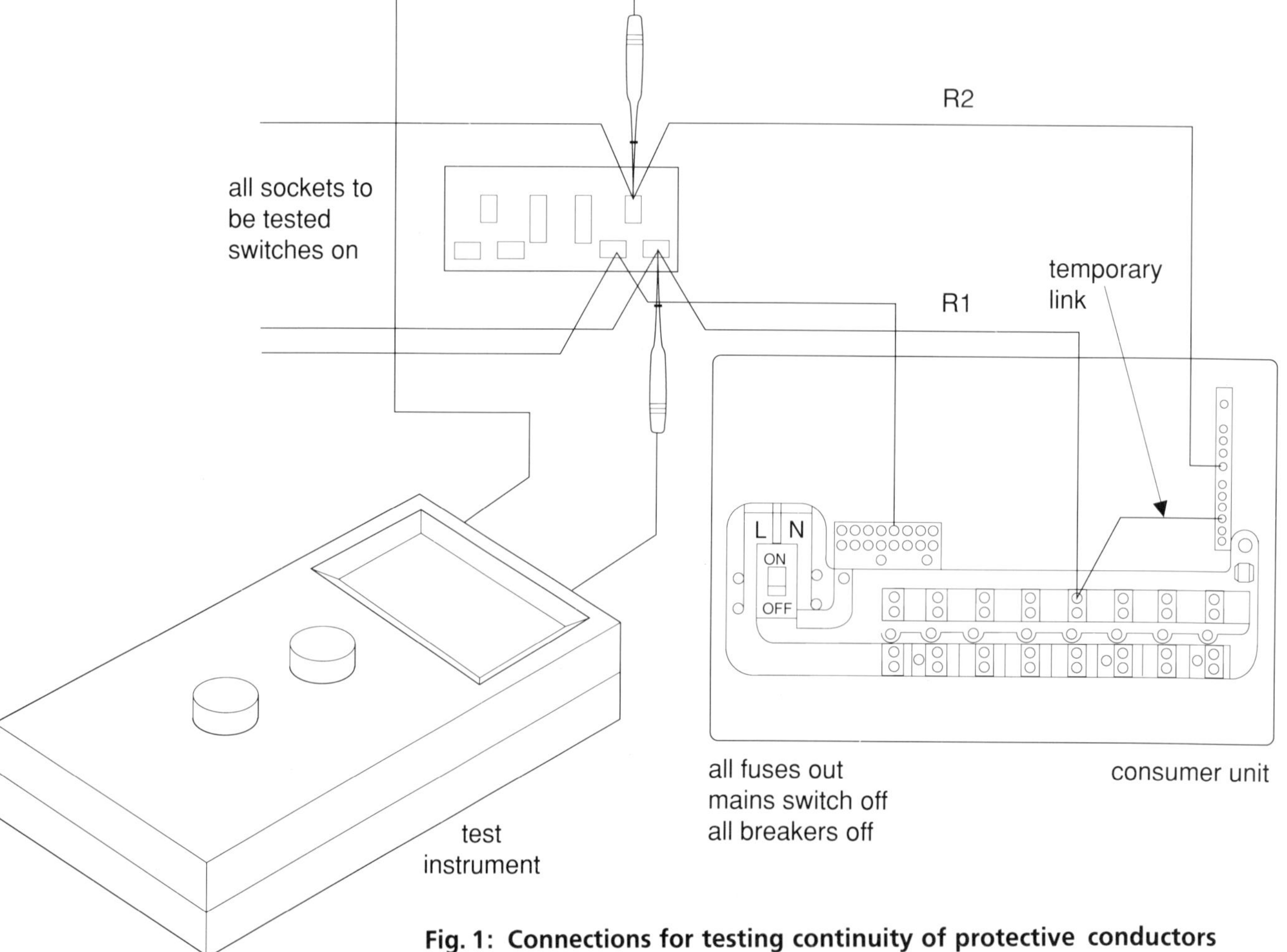

Fig. 1: Connections for testing continuity of protective conductors

Notes: Where the alternative method to Table 41A of BS 7671 has been used for sizing the protective conductor, it will be necessary to make a separate measurement of the protective conductor (R_2) using Test method 2.   413-02-12

It should also be recognised that these methods of test can only be applied simply to an 'all insulated' installation. Installations incorporating steel conduit, steel trunking, micc and pvc/swa cables will introduce parallel paths to protective conductors. Similarly, luminaires fitted in grid ceilings and suspended from steel structures in buildings will create parallel paths.

In this situation unless a plug and socket arrangement has been incorporated in the lighting system by the designer the (R_1+R_2) test needs to be carried out prior to fixing accessories and bonding straps to the metal enclosures and finally connecting protective conductors to luminaires. Under these circumstances some of the requirements may have to be visually inspected after the test has been completed. This consideration requires tests to be performed during the process of erecting an installation, in addition to tests at the completion stage. 713-02

Where ferrous enclosures have been used as the protective conductors, eg conduit, trunking, steel-wire armouring etc the following special procedure should be followed:

(i) perform the standard ohmmeter test using the appropriate test method described above.

Instrument: Use a low resistance ohmmeter for this test.

(ii) inspect the enclosure along its length to verify its integrity

(iii) if it is felt by the inspecting engineer that there may be grounds to question the soundness of this conductor, a further test should be performed using a phase-earth loop impedance tester after the connection of the supply.

Instrument: Use a loop impedance tester for this test. Refer to Section 16.

(iv) if, following this further test, the inspecting engineer still feels that there may be grounds to question the soundness of this conductor, a further test may be performed using an ac ohmmeter which has a test voltage not exceeding 50 V and can provide a test current approaching 1.5 times the design current of the circuit, excepting that it need not exceed 25 A.

Care needs to be taken when using such high current testers, as sparking can occur at a faulty joint. This test should not be carried out if this could be dangerous.

Instrument: Use a high-current low-impedance ohmmeter for this test.

8.2 Continuity of ring final circuit conductors 713

A test is required to verify the continuity of each conductor including the circuit protective conductor (cpc) of every ring final circuit. The test results should establish that the ring is complete and has not been interconnected.

713-03

It may be possible, as an alternative, in order to establish that no interconnecting multiple loops have been made in a ring circuit, for the inspector to check visually each conductor throughout its entire length.

Instrument: Use a low-resistance ohmmeter for these tests. Refer to Section 16.

Test method

The phase, neutral and protective conductors are identified and their resistances are measured separately (see Fig. 2a). The end to end resistance of the protective conductor r_2 is divided by 4 to give R_2 and is recorded on the installation schedule form WR4. A finite reading confirms that there is no open circuit on the ring conductors under test. The resistance values obtained should be the same (within 0.05 ohms) if the conductors are the same size. If the protective conductor has a reduced csa the resistance of the protective conductor loop will be proportionally higher than that of the phase or neutral loop eg 1.67 times for 2.5/1.5 mm^2 cable. If these relationships are not achieved then either the conductors are incorrectly identified or there is a loose connection at one of the accessories.

The phase and neutral conductors are then connected together so that the outgoing phase conductor is connected to the returning neutral conductor and vice versa (see Fig. 2b). The resistance between phase and neutral conductors is then measured at each socket-outlet. The readings obtained from those sockets wired into the ring will be substantially the same and the value will be approximately half the resistance of the phase or the neutral loop resistance. Any sockets wired as spurs will have a proportionally higher resistance value corresponding to the length of the spur cable.

The latter exercise is then repeated with the phase and cpc cross connected as before (see Fig. 2c). The resistance between phase and earth is then measured at each socket. The highest value recorded represents the maximum $(R_1 + R_2)$ of the circuit and is recorded on form WR4 and can be used to determine the earth loop impedance (Z_S) of the circuit to verify compliance with the loop impedance requirements of the Regulations (See 14.2).

$(R_1 + R_2)$ is equal to $(r_1 + r_2)/4$, where r_1 and r_2 are the end to end readings taken at the start of the test for ring continuity.

Note: Where single-core cables are used, special care should be taken to verify that the phase and neutral conductors of opposite ends of the ring circuit are bridged together. An error in this respect will be apparent from the readings taken at the socket-outlets, progressively increasing in value as readings are taken towards the midpoint of the ring, then progressively decreasing towards the other end of the ring.

This sequence of tests verifies the polarity of each socket and dispenses with the need for a separate protective conductor continuity test (See 8.1).

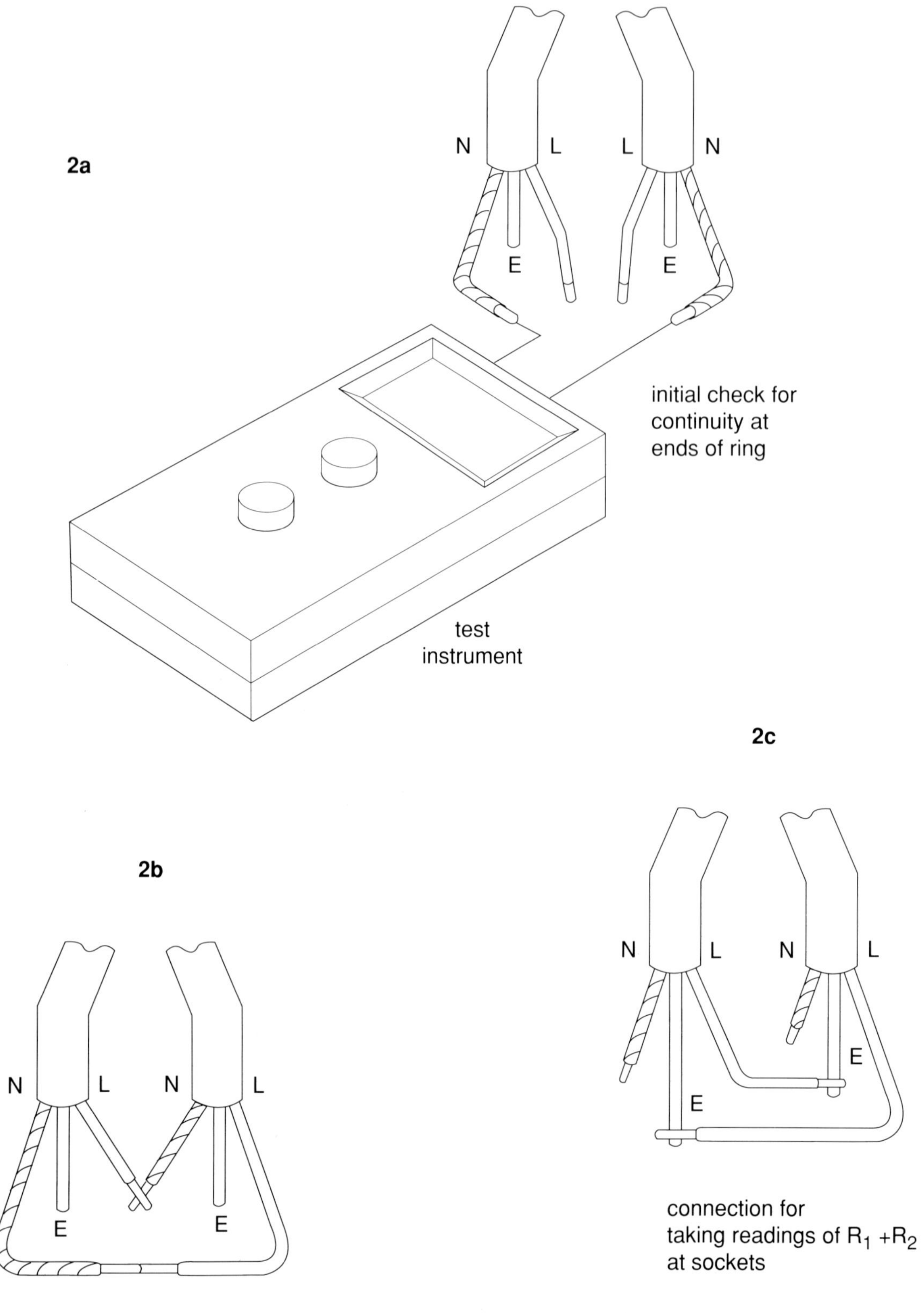

Fig.2: Connections for testing continuity of ring final circuit conductors

Section 9 — Insulation Resistance

9.1 Insulation resistance 713

These tests are to verify that for compliance with the Wiring Regulations the insulation of conductors and electrical accessories and equipment is satisfactory and that electrical conductors or protective conductors are not short-circuited, or show a low insulation resistance (which would indicate defective insulation).

713-04-01 to 04

Before testing check that:

(i) pilot or indicator lamps, and capacitors are disconnected from circuits to avoid an inaccurate test value being obtained

(ii) if test 1 is to be used voltage-sensitive electronic equipment such as dimmer switches, touch switches, delay timers, power controllers, electronic starters for fluorescent lamps, emergency lighting, rcds etc are disconnected so that they are not subjected to the test voltage

(iii) there is no electrical connection between any phase or neutral conductor and earth.

If circuits contain voltage sensitive devices, test 2, from protective earth to (phase and neutral) connected together, may be appropriate if vulnerable equipment is not to be disconnected. Further precautions may also be necessary to avoid damage to some electronic devices.

713-04-06

Instrument: Use an insulation resistance tester for these tests. Refer to Section 16.

Insulation resistance tests should be carried out using the appropriate dc test voltage specified in Table 71A of BS 7671. The installation will be deemed to conform with the Regulations if the main switchboard, and each distribution circuit tested separately with all its final circuits connected, but with current using equipment disconnected, has an insulation resistance not less than that specified in Table 71A. Table 71A of BS 7671 is reproduced as Table 9.1.

Simple installations that contain no distribution circuits should be tested as a whole.

The tests should be carried out with main switch off, all fuses in place, switches and circuit-breakers closed, lamps removed, and fluorescent and discharge luminaires and other equipment disconnected. Where the removal of lamps and/or the disconnection of current-using equipment is impracticable, the local switches controlling such lamps and/or equipment should be open.

TABLE 9.1
Minimum values of insulation resistance

Circuit nominal voltage (V)	Test Voltage dc (V)	Minimum insulation resistance (MΩ)
SELV and PELV	250	0.25
Up to and including 500 V with the exception of the above systems	500	0.5
Above 500 V	1000	1.0

To perform the test in a complex installation it may need to be sub-divided into its component parts.

Although an insulation resistance value of not less than 0.5 megohms complies with the Regulations, where an insulation resistance of less than 2 megohms is recorded, the possibility of a latent defect exists. If this is the case, each circuit should be separately tested and its measured insulation resistance should be greater than 2 megohms.

Test 1

Insulation resistance between live conductors.

Single-phase

Test between the phase and neutral conductors at appropriate switchboard.

Three-phase

Make a series of tests between live conductors in turn at the appropriate switchboard as follows:

 (i) between phase 1 and (phase 2, phase 3 and neutral) grouped

 (ii) between phase 2 and (phase 3 and neutral) grouped

 (iii) between phase 3 and neutral.

Where it is not possible to group conductors in this way, conductors may be tested singly. Resistance readings obtained should not be less than the minimum values referred to in Table 9.1.

See Fig. 3. for Test method connections, for the testing of individual circuits in which protective conductors to switches have been omitted for clarity.

Test 2

Insulation resistance from earth to phase and neutral connected together.

Single-phase

Test between the phase and neutral conductors and earth at the appropriate distribution board.

Where any circuits contain two-way switching the two-way switches will require to be operated and another insulation resistance test carried out, including the two-way strapping wire which was not previously included in the test.

Three-phase

Measure between all phase conductors and neutral bunched together, and earth. Where a low reading is obtained (less than 2 MΩ) it may be necessary to separately test each conductor to earth.

Resistance readings obtained should not be less than the minimum values referred to in Table 9.1.

See Fig. 4 for Test method connections.

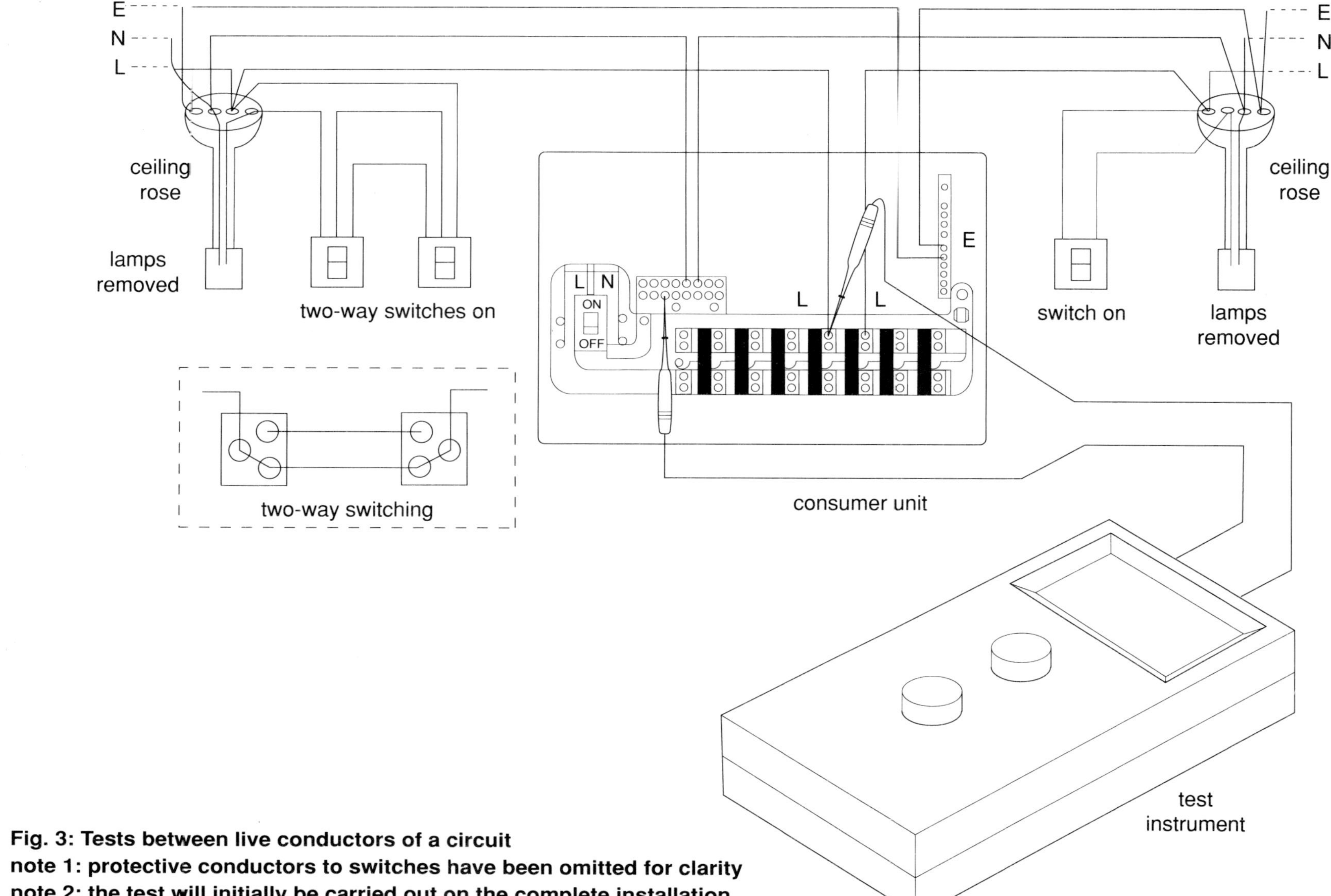

Fig. 3: Tests between live conductors of a circuit
note 1: protective conductors to switches have been omitted for clarity
note 2: the test will initially be carried out on the complete installation,
see paragraph 9.1.

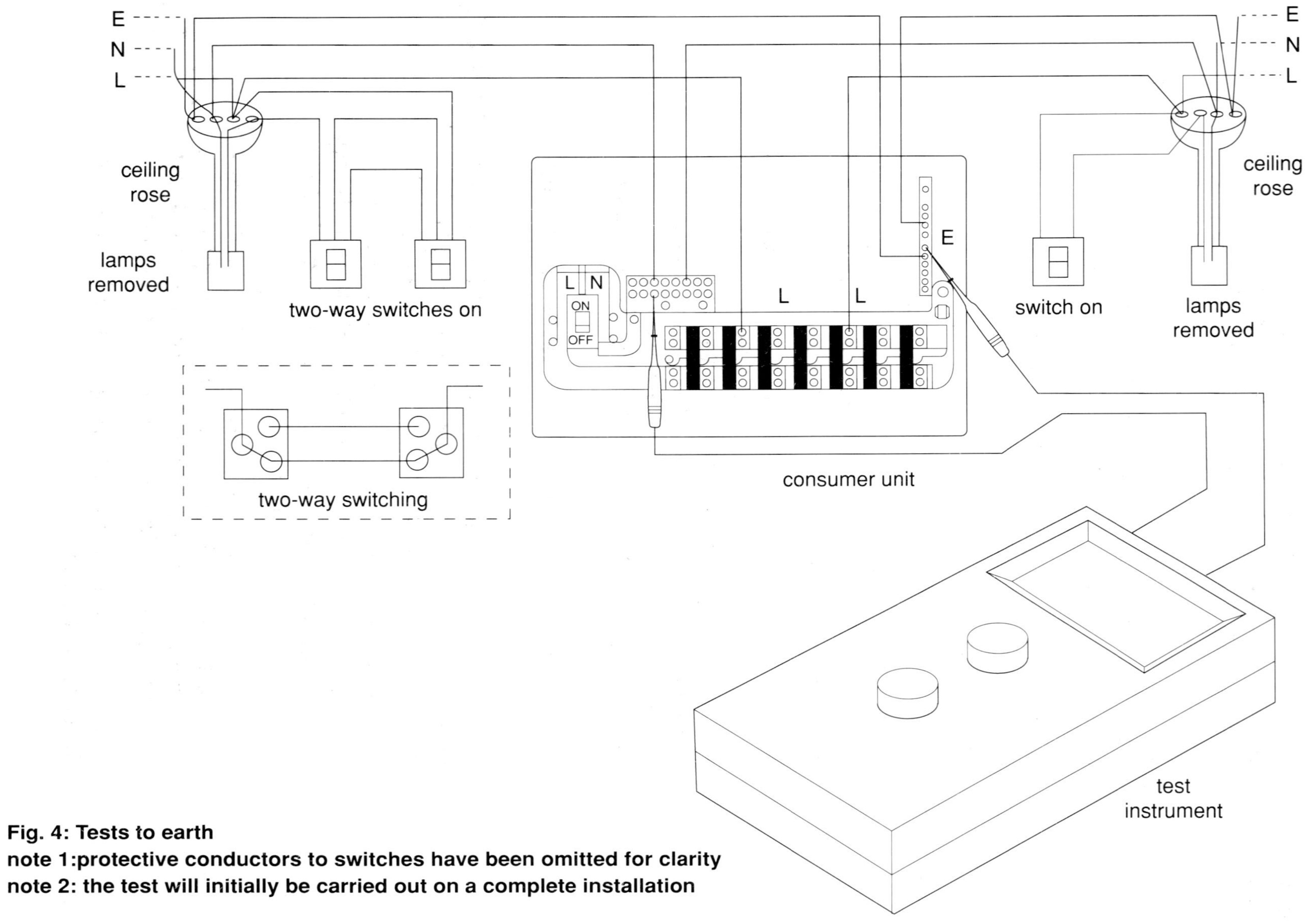

Fig. 4: Tests to earth
note 1:protective conductors to switches have been omitted for clarity
note 2: the test will initially be carried out on a complete installation

9.2 Site applied insulation
412
713

These tests are applicable only where insulation is applied during erection. They involve the use of high voltage and great care is necessary to avoid danger.   412-02

When protection against direct contact is afforded by insulation which has been applied to live parts of equipment during erection on site, a test should be made to check that the insulation is capable of withstanding an applied test voltage equivalent to that specified in the British Standard for similar factory-built equipment.   713-05-01

The test voltage is applied between the live conductors connected together, and metallic foil wrapped closely around all surfaces of the insulation. The test voltage and duration must be in line with the appropriate British Standard specification.

Where there is no such British Standard, the test should be applied using a test voltage of 3750 V ac. This test voltage should be at supply frequency, and should be applied to the insulation for a duration of 1 minute. The insulation will be deemed to be satisfactory if no flashover breakdown occurs during the period of test.

Instrument: Use an applied voltage tester for this test. Refer to Section 16.

Where protection against indirect contact is provided by supplementary insulation applied to equipment during erection, a test should be made to verify that: 713-05-02

  (i)   the insulating enclosure affords a degree of protection not less than IP2X or IPXXB. For details of test methods refer to Section 11 (Protection by barriers or enclosures)

  (ii)   the insulating enclosure is capable of withstanding, without breakdown or flashover, an applied test voltage equivalent to that specified in the British Standard for similar factory-built equipment.

Instrument: Use an applied voltage tester for this test. Refer to Section 16.

Section 10— Protection by Separation of Circuits

10.1 Care in use of high voltage

When tests involve the use of high voltage great care is necessary to avoid danger. HS(G)13 should be consulted. In the example of 3750 V ac, for the additional withstand test in paragraph 9.2, the voltage is taken from BS 3456: Part 101: 1987 Table 5. The test voltages related to BS 3535 are for type testing. For transformers up to and including 1000 VA, in production tests, the electric strength test voltage is only applied for 2 seconds. Withstand tests are applied initially at half volts and rapidly raised to full voltage.

Note: Verification of measures for protection against indirect contact is not specified in the Regulations.

10.2 SELV
713-06-01
713-06-02

The source of voltage must be inspected to confirm compliance with the Wiring Regulations. — 411-02-02 / 411-02-04

The live parts of the SELV (separated extra-low voltage) circuits must be tested to verify electrical separation in accordance with the Wiring Regulations. — 412-02-05

This is achieved by testing between the live conductors of each SELV circuit connected together and the live conductors of any higher voltage circuits connected together.

The first test applied to this arrangement is an insulation resistance test in accordance with Table 9.1. — Table 71A

Where the circuit is supplied from a safety source complying with the Wiring Regulations, the test is made at 500 V dc and the required insulation resistance value is 5 megohms. The reading is taken after the voltage has been applied for 1 minute. — 411-02-02 / 411-02-04

Instrument: Use an insulation resistance tester for this test. Refer to Section 16.

The SELV circuit conductors should be inspected to verify compliance with the Wiring Regulations. 411-02-06 (i)(ii)(iv)

713-06-02 To demonstrate compliance with the Wiring Regulations an additional insulation resistance test should be applied at 500 V dc between the live conductors of the SELV circuit and the protective conductor of the higher voltage present. The insulation resistance should be not less than 5 megohms after 1 minute.

Instrument: Use an insulation resistance tester for this test. Refer to Section 16.

Compliance with Regulation 411-02-08 will have been verified by the inspection required by Regulation 411-02-06 (iii). This should be confirmed. 411-02-06 411-02-08

The exposed-conductive-parts should be inspected to confirm compliance with Regulation 411-02-09. 411-02-09

The source of the SELV supply should be inspected for conformity with Regulation 411-02-02 and if necessary the voltage should be measured to ensure that the voltage does not exceed 50 V ac or 120 V dc. If the voltage exceeds 25 V ac or 70 V dc (60 V dc ripple-free), the means of protection against direct contact should be verified. This may be by either: 411-02-02

(i) inspection of the barriers or enclosures to ensure they offer a degree of protection not less than IP2X (IPXXB) or IP4X as required by Regulation 412-03, or

(ii) inspection of the insulation and applying a 500 V dc insulation resistance test between the live conductors of the SELV circuit and the accessible insulation. The insulation resistance should be not less than 5.0 megohms the reading being taken after applying the voltage for 1 minute. 411-02-03 Table 71A

10.3 PELV **713-06-01**	PELV (protective extra-low voltage) installations are inspected and tested as for SELV installations except that an insulation test is not made between PELV circuits and earth.	471-14
713-06-03	(PELV systems may include a protective conductor connected to the protective conductor of the primary circuit and to one point on the secondary of the ELV source. Protection against indirect contact in the secondary circuit is dependent upon the primary circuit protection).	
10.4 Functional **extra-low** **voltage** **713-06-05**	Extra-low voltage circuits not meeting the requirements for SELV or PELV are inspected and tested as low voltage circuits.	471-14-03 471-14-06
10.5 Electrical **separation** **713-06-04**	The source of supply should be inspected to confirm compliance with the Regulations. In addition, should any doubt exist, the voltage should be measured to verify it does not exceed 500 V.	713-06-02 413-06-02 (iv)
	Where the source of supply does not comply with Regulation 413-06-02(i), compliance with Regulation 413-03 must be verified. Where this cannot be done or doubt exists, the insulation between live parts of the separated circuit and any other conductor, or to earth, must be tested. This test should be performed at a voltage of 500 V dc and the insulation resistance should be not less than 0.5 megohm.	413-06-02 (iii)(b) 413-03 Table 71A
	Instrument: Use an insulation resistance tester for this test. Refer to Section 16.	
	The live parts of the separated circuit must be tested to ensure that they are electrically separate from other circuits. This is achieved by testing between the live conductors of the separated circuit connected together and the conductors of any other circuit strapped together.	413-06-02 (iii) (iv)
	The first test applied to this arrangement is an insulation resistance test at 500 V dc. The insulation resistance should be not less than 0.5 megohm.	

Instrument: Use an insulation resistance tester for this test. Refer to Section 16.

A separate wiring system is preferred for electrical separation. If multicore cables or insulated cables in insulated conduit are used, all cables must be insulated to the highest voltage present and each cable shall be protected against overload and fault current.

413-06-03 (iii)

A 500 V dc insulation resistance test is performed between the exposed-conductive-parts of any item of equipment connected, and the protective conductor or exposed-conductive-parts of any other circuit, to confirm compliance with the Regulations. The insulation resistance should not be less than 0.5 megohm.

413-06-03 (iv)

Instrument: Use an insulation resistance tester for this test. Refer to Section 16.

If the separated circuit supplies more than one item of equipment, the requirements of the Regulations shall be verified as follows:

413-06-05

(i) apply a continuity test between all exposed-conductive-parts of the separated circuit to ensure they are bonded together. This equipotential bonding is then subjected to a 500 V dc insulation resistance test between the protective conductor or exposed-conductive-parts of other circuits, or to extraneous-conductive-parts. The insulation resistance should not be less than 0.5 megohm.

Instrument: Use an insulation resistance tester for this test. Refer to Section 16.

(ii) all socket-outlets must be inspected to ensure that protective conductor contact is made to the equipotential bonding conductor

(iii) all flexible cables other than those feeding Class II equipment must be inspected to ensure that they embody a protective conductor for use as an equipotential bonding conductor

(iv) operation of the protective device must be   Table 41A
verified by measurement of the fault loop   and 41B
impedances to the various pieces of equipment
connected, with reference to the type and rating of
the protective device for the separated circuit. If
protection is provided by overcurrent devices, the
appropriate value of loop impedance given by
Regulation 413-02-08 (Tables 41B, C and D for
230 V systems) shall be determined.

Instrument: Use a loop impedance tester for this test.
Refer to Section 16.

Section 11 — Protection by Barriers or Enclosures

11.1 Protection by barriers or enclosures provided during erection 713 412 413

This test is not applicable to barriers or enclosures provided in factory-built equipment, but only to those provided on site during the course of assembly or erection. Where during erection, an enclosure or barrier is provided, to afford protection from direct contact a degree of protection not less than IP2X or IPXXB is required. Readily accessible horizontal top surfaces shall have a degree of protection equivalent to IP4X.

713-07
412-03-01
412-03-02
412-03-04

The degree of protection afforded by IP2X is defined in BS EN 60529 as protection against the entry of 'Fingers or similar objects not exceeding 80 mm in length. Solid objects exceeding 12 mm in diameter'. The test is made with a metallic standard test finger.

412-03-01

Both joints of the finger may be bent through 90° with respect to the axis of the finger, but in one and the same direction only. The finger is pushed without undue force (not more than 10 N) against any openings in the enclosure and, if it enters, it is placed in every possible position.

A low voltage supply (of not less than 40 V and not exceeding 50 V for safety reasons) in series with a suitable lamp is connected between the test finger and the live parts inside the enclosure. Conducting parts covered only with varnish or paint or protected by oxidation or by a similar process, shall be covered with a metal foil electrically connected to those parts which are normally live in service.

The protection is satisfactory if the lamp does not light.

The degree of protection afforded by IP4X is defined in BS EN 60529 as protection against the entry of 'Wires or strips of thickness greater than 1.0 mm, and solid objects exceeding 1.0 mm in diameter'.

412-03-02

The test is made with a straight rigid steel wire of 1 mm + 0.05/0 mm diameter applied with a force of 1 N ± 10 per cent. The end of the wire shall be free from burrs, and at right angles to its length.

The protection is satisfactory if the wire cannot enter the enclosure.

Note: Reference should be made to the appropriate product standard or BS EN 60529 for a fuller description of the degrees of protection, details of the standard test finger and other aspects of the tests.

Section 12 — Protection by a Non-conducting Location

12.1 Insulation of non-conducting floors and walls
413
713

Where protection against indirect contact is provided by a non-conducting location the following should be verified:

413-04
713-08

(i) exposed-conductive-parts should be arranged so that under normal circumstances a person will not come into simultaneous contact:

413-04-02

with two exposed-conductive-parts or

with an exposed-conductive-part and any extraneous-conductive-part.

(ii) in the non-conducting location there must be no protective conductors

(iii) any socket-outlets installed in a non-conducting location must not incorporate an earthing contact

413-04-03

(iv) the resistance of insulating floors and walls to the main protective conductor of the installation should be tested at not less than three points on each relevant surface, one of which should be not less than 1 m and not more than 1.2 m from any extraneous-conductive-part, eg pipes, in the location. 12.2 describes a method of measuring the insulation resistance of floors and walls.

413-04-04
713-08-01

If at any point the resistance is less than the specified value, the floors and walls are extraneous-conductive-parts as detailed in the Regulations.

413-04-04

If any extraneous-conductive-part is insulated it should be tested and must be capable of withstanding at least 2 kV rms ac without breakdown, and should not pass a leakage current exceeding 1 mA in normal use.

413-04-07
713-08-02

Note: The following test involves the use of high voltage and adequate precautions should be taken to prevent danger. Consult HS(G)13.

This testing is done by applying metallic foil closely around all the insulation, and then firstly applying 2 kV ac between this foil and the main protective conductor for a duration of 1 minute. The insulation will be deemed to be satisfactory if no flashover or breakdown occurs during the test. This test is then followed by an insulation resistance test at 500 V dc across the same connections. The insulation resistance should not be less than 0.5 megohm.

413-04-07

Instrument: Use a high output applied voltage tester for this test. Refer to Section 16 for additional precautions.

12.2 Measuring insulation resistance of floors and walls

A magneto-ohmmeter or battery-powered insulation tester providing a no-load voltage of approximately 500 V (or 1000 V if the rated voltage of the installation exceeds 500 V) is used as a dc source.

The resistance is measured between the test electrode and a protective conductor of the installation.

The test electrodes may be either of the following types. In case of dispute, the use of test electrode 1 is the reference method.

Note: It is recommended that the test be made before the application of the surface treatment (varnishes, paints and similar products).

Test electrode 1

The electrode comprises a square metallic plate with sides 250 mm and a square of damped water absorbent paper or cloth from which surplus water has been removed with side approximately 270 mm which is placed between the metal plate and the surface being tested.

During the measurement a force of approximately 750 N (12 stone in weight) or 250 N is applied on the plate, in the case of floors or of walls respectively.

Test electrode 2

The test electrode comprises a metallic tripod of which the parts resting on the floor form the points of an equilateral triangle. Each supporting part is provided with a flexible base ensuring, when loaded, close contact with the surface being tested over an area of approximately 900 mm^2 and presenting a resistance of less than 5000 Ω.

Before measurements are made, the surface being tested is moistened or covered with a damp cloth. While measurements are being made a force of approximately 750 N or of 250 N is applied to the tripod, in the case of floors or of walls respectively.

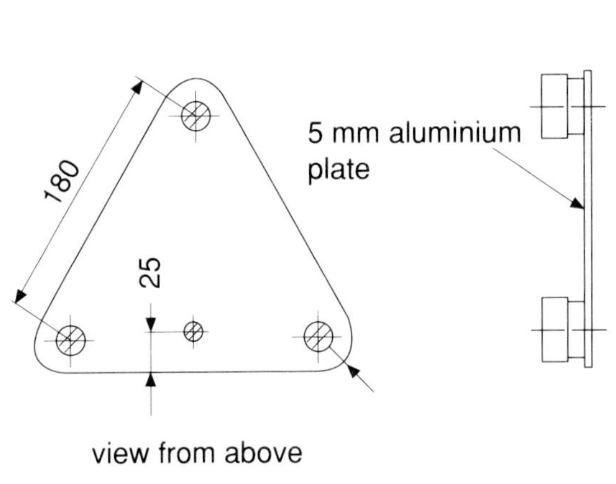

view from above

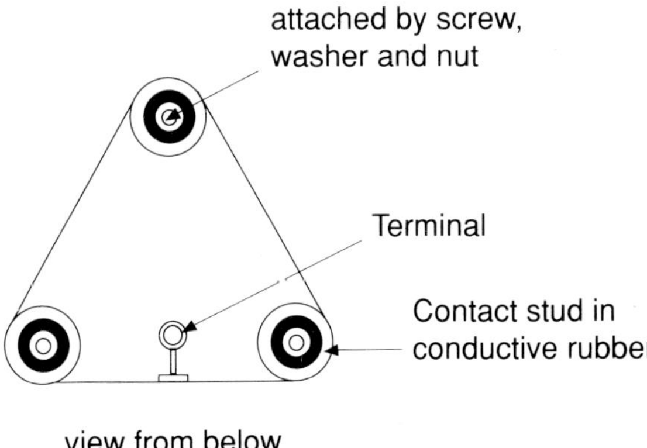

view from below

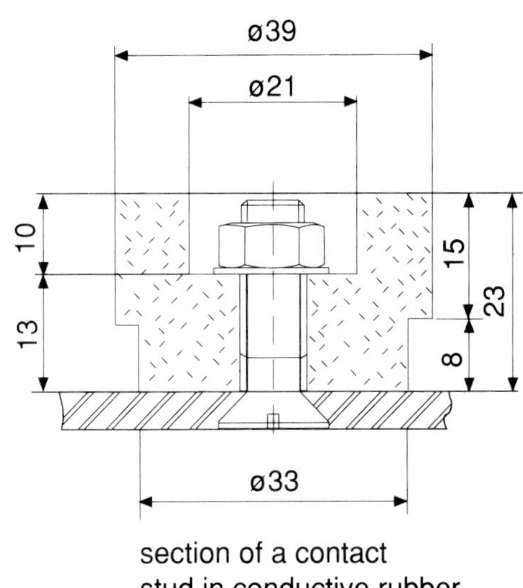

section of a contact stud in conductive rubber

Fig.5: Test electrode 2

Section 13 — Polarity Tests

A test needs to be performed to check the polarity of all circuits. This must be done before connection to the supply, with either an ohmmeter, or the continuity range of an insulation and continuity tester.

713-09

Instrument: Use a low-resistance ohmmeter for these tests. Refer to Section 16.

The method of test is the same as described for Test method 1 paragraph 8.1 for checking the continuity of protective conductors.

713-02
713-03

For ring circuits, if the tests required by Regulation 713-03 have been carried out, the correct connections of phase, neutral and cpc conductors will have been verified and no further testing is required.

713-03

For radial circuits the $(R_1 + R_2)$ measurements, made as in Test method 1 paragraph 8.1, should be made at each point.

For lighting circuits it is necessary to check that all fuses and single-pole switches are connected to the phase conductor. The line connection in socket-outlets and the centre contact of screw-type lampholders must all be connected to the phase conductor. This can be checked as shown in Fig. 6.

See Fig. 6 for Test method connections.

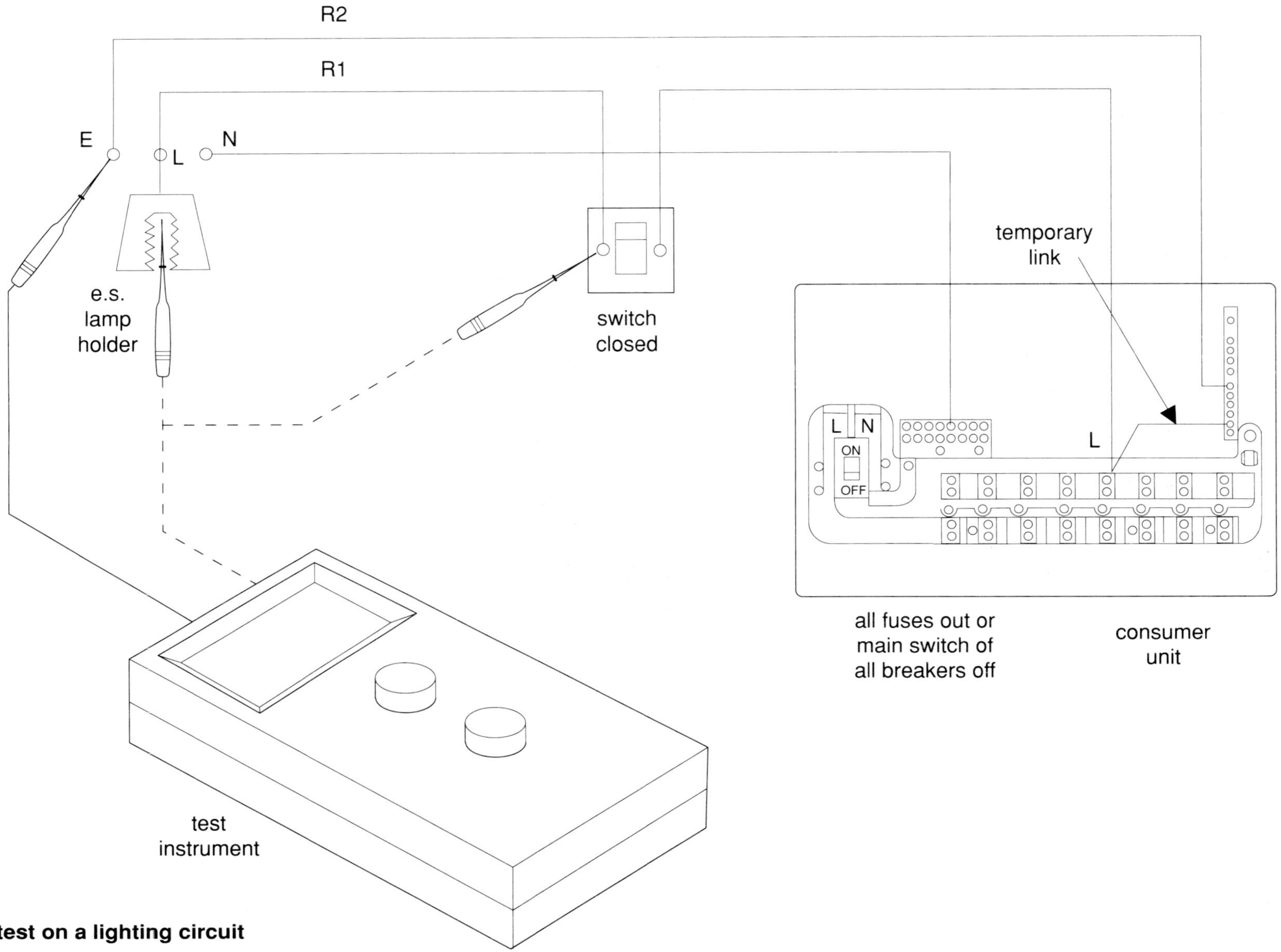

Fig. 6: Polarity test on a lighting circuit

Section 14 — Earthing

14.1 Earth electrode resistance 713 413

After an earth electrode has been installed it is necessary to verify that the resistance meets the conditions of the Wiring Regulations for TT and IT installations.

713-11
413-02-20
413-02-23

Instrument: Use an earth electrode resistance tester for this test. Refer to Section 16.

(i) *Test method 1*

The test requires the use of two test spikes (electrodes), and is carried out in the following manner.

Connection to the earth electrode is made using terminals C1 and P1 of a four-terminal earth tester having first ensured that the earth lead is disconnected. To exclude the resistance of the test leads from the resistance reading, individual leads should be taken from these terminals and connected separately to the electrode. If the test lead resistance is insignificant, the two terminals may be short-circuited at the tester and connection made with a single test lead, the same being true if using a three-terminal tester. Connection to the temporary spikes is made as shown in Fig. 7.

The distance between the test spikes is important. If they are too close together, their resistance areas will overlap. In general, reliable results may be expected if the distance between the electrode under test and the current spike is at least ten times the maximum dimension of the electrode system, eg 30 m for a 3 m long electrode.

Three readings are taken: with the potential spike
initially midway between the electrode and current
spike, secondly at a position 10 per cent of the
electrode-to-current spike distance back towards the
electrode, and finally at a position 10 per cent of the
distance towards the current spike.

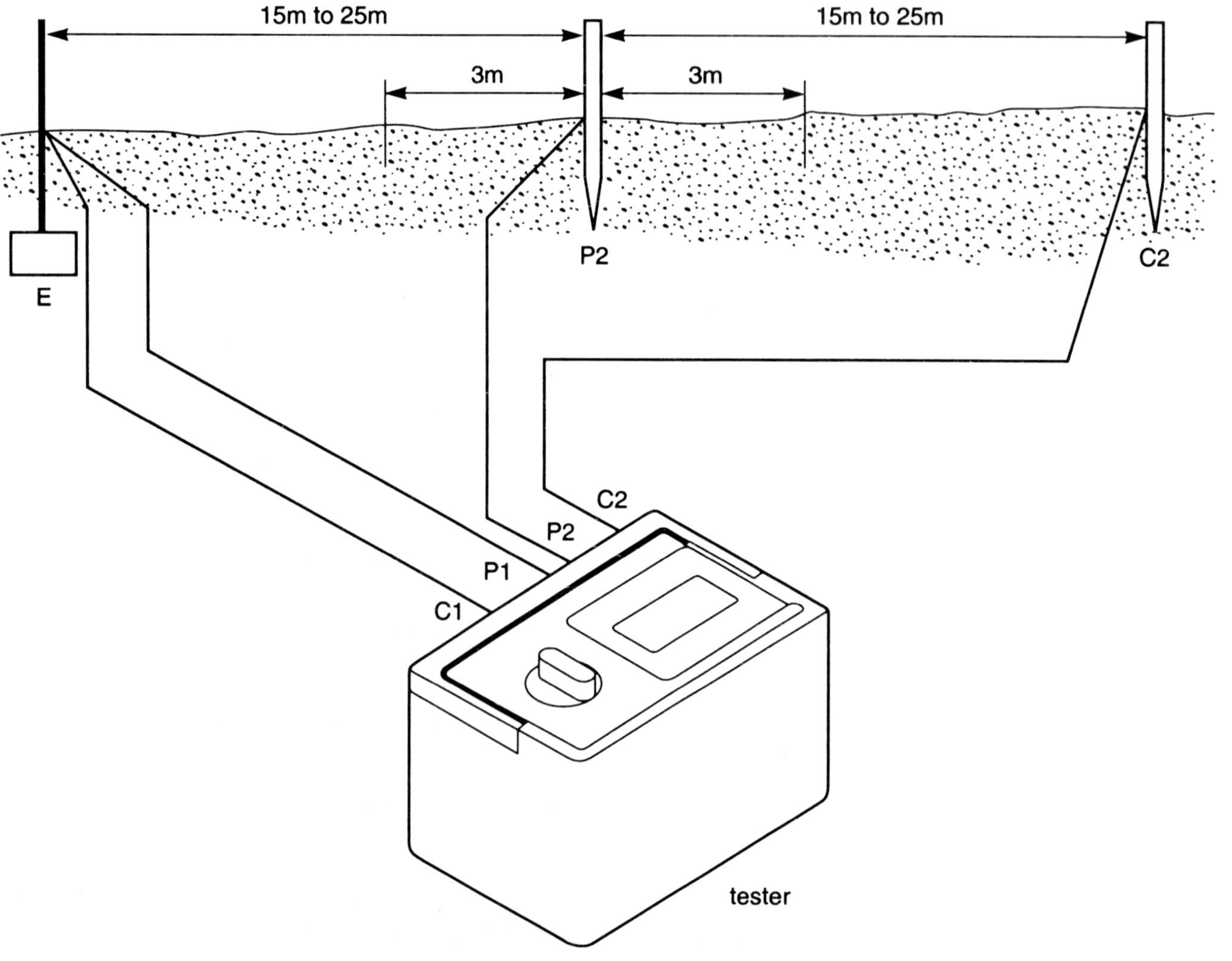

Fig.7: Earth Electrode Test

By comparing the three readings, a percentage
deviation can be determined. This is calculated by
taking the average of the three readings, finding the
maximum deviation of the readings from this average
in ohms, and expressing this as a percentage of the
average.

The accuracy of the measurement using this technique is typically 1.2 times the percentage deviation of the readings. It is difficult to aim for a measurement accuracy better than 2 per cent, and inadvisable to accept readings that differ by more than 5 per cent. To improve the accuracy of the measurement to acceptable levels, the test must be repeated with a larger separation between the electrode and the current spike.

The instrument output current may be ac or reversed dc to overcome electrolytic effects. Because these testers employ phase-sensitive detectors (psd), the errors associated with stray currents are eliminated.

The instrument should be capable of checking that the resistance of the temporary spikes used for testing are within the accuracy limits stated in the instrument specification. This may be achieved by an indicator provided on the instrument, or the instrument should have a sufficiently high upper range to enable a discrete test to be performed on the spikes.

If the temporary spike resistance is too high, measures to reduce the resistance will be necessary, such as driving the spikes deeper into the ground or watering with brine to improve contact resistance. In no circumstances should these techniques be used to temporarily reduce the resistance of the earth electrode under test.

After completion of the testing ensure that the earth lead is reconnected.

(ii)   *Test method 2 (alternative for rcd protected TT installations)*

If the electrode under test is being used in conjunction with a residual current device protecting a TT installation, the following method of test may be applied as an alternative to the earth electrode resistance test described above. This test method is relatively tolerant to inaccuracies in measurement.

413-02-16

Before this test is undertaken, the earthing conductor to the earth electrode should be disconnected at the main earthing terminal to ensure that all the test current passes through the earth electrode alone.

A loop impedance tester is connected between the
phase conductor at the source of the TT installation
and the earth electrode, and a test performed. The
impedance reading taken is treated as the electrode
resistance and is then added to the resistance of the
protective conductor for the protected circuits. The
product of this value in ohms and the residual
operating current of the rcd in amperes must not
exceed 50 V.

413-02-16

If this value does exceed 50 V, Test method 1 should
be employed to check the actual value of the
electrode resistance.

AFTER THE TEST, ENSURE THAT EQUIPOTENTIAL
BONDS ARE RECONNECTED.

Instrument: Use a loop-impedance tester for this test.
Refer to Section 16.

Test results

The test results required will depend upon the
application of the electrode and reference should be
made to one or more of the following:

(i)   for TT and IT systems to comply with the
disconnection times of the Regulations the
product of the earth electrode resistance
and operating current of the protection device
must not exceed 50 V for normal locations

413-02-12
413-02-20
413-02-23
Part 6

(ii) when protection is afforded by a residual current
device in an installation which is part of a TN or
TT system, the product of the rated residual
operating current in amperes and the earth
fault loop impedance in ohms should not
exceed 50 V for normal locations. It may be
reduced in special locations

413-02-16

Part 6

(iii) with regard to higher resistance values British
Standard 7430 Code of practice for earthing
suggests that values of electrode resistance of
200 ohms or above may be unstable.

14.2 Earth fault loop impedance

The earth fault impedance Z_S should be determined at the farthest point of each circuit including socket-outlets, lighting points, distribution circuits and any other fixed equipment. The value obtained, after adjustment to take into account the heating effects of load current, should not exceed that detailed in Table 41B or Table 41D, or should not exceed a value which might prevent conformity with the Regulations for rcd protected circuits.

713-10
413-02-16
Table 41B
Table 41D

The earth fault current loop comprises the following parts, starting at the point of fault on the phase to earth loop:

(i) circuit protective conductor

(ii) the main earthing terminal and earthing conductors

(iii) for TN systems the metallic return path (or in the case of TT and IT systems the earth return path)

(iv) the path through the earth and neutral point of the transformer

(v) the transformer winding and the phase conductor from the transformer to the point of fault.

Test method to measure Z_e

Z_e is measured using a phase earth fault loop impedance tester at the source of the installation supply.

The impedance measurement is made between the main phase supply and the main means of earthing with the main switch open or with all the circuits isolated. The means of earthing will be isolated from the installation earthed equipotential bonding for the duration of the test. Care should be taken to avoid any shock hazard to the testing personnel and other persons on the site both whilst establishing contact, and performing the test. Ensure that the earth connection has been replaced before reclosing the main switch.

542-04-02

See Fig. 8 for Test method connections.

Instrument: Use a loop impedance tester for this test. Refer to Section 16.

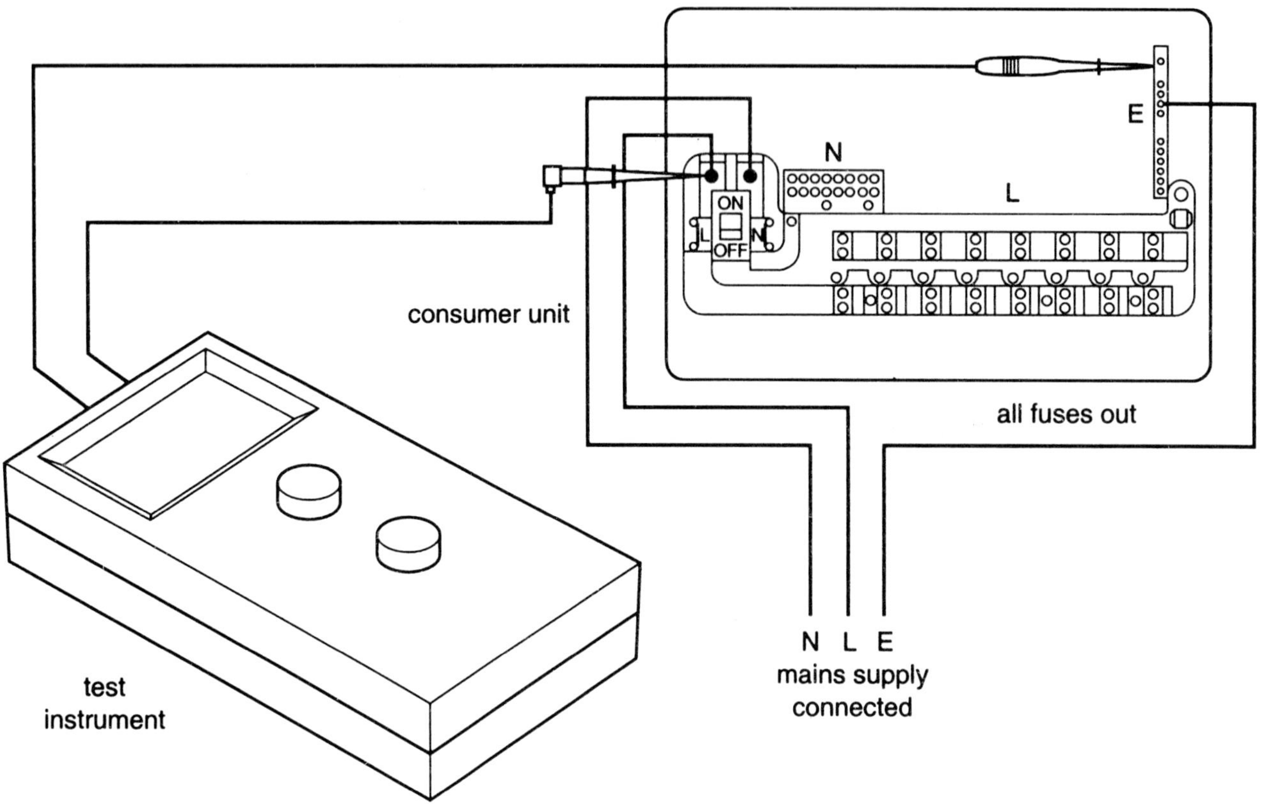

Fig.8: Test of Z_e at the source of supply

The diagram shows the test probes connected to phase and earth only. Three-wire loop impedance testers require a connection to the neutral for the instrument to operate.

Measurement of (R_1+R_2)

Whilst testing the continuity of protective conductors of radial circuits, or whilst testing the continuity of ring final circuits, the value of (R_1+R_2) is measured at ambient temperature.

The measured value of (R_1+R_2) for the final circuit should be added to the appropriate values of (R_1+R_2) for any distribution circuits.

An alternative means of determining (R_1+R_2) is to measure the loop impedance value at the extremity of the final circuit, taking care to use the correct phase supply and protective conductor return for the circuit. This loop impedance less the Z_e value measured earlier can be taken as the value of (R_1+R_2).

These (R_1+R_2) values will be used to determine the value of Z_S of the circuit under fault conditions, but must be first corrected for conductor temperature. This requires the use of correction factors dependent on the type of cable used in the circuit, and ambient temperature at the time of continuity testing.

Temperature correction

The total (R_1+R_2) value must first be multiplied by a factor which is dependent on the ambient temperature at the time that the appropriate continuity test was performed. These correction factors for test ambient temperatures other than 20°C are:

434-03-03
543-01-03

TABLE 14A

Ambient temperature	Correction factor (Ca)
5°C	1.06
10°C	1.04
15°C	1.02
25°C	0.98

The resulting value in ohms must then be multiplied by a correction factor applicable to the type of cable insulation used on the circuit conductor, as this determines the maximum operating temperature of the cable. The multipliers are given in Table 14B.

TABLE 14B

Insulation Material		pvc	85°C rubber	90°C thermosetting
Multiplier	54B	1.04	1.04	1.04
	54C	1.20	1.26	1.28

The multipliers are based on the simplified formula given in BS 6360 for both copper and aluminium conductors namely that the resistance-temperature coefficient is 0.004 per °C at 20°C.

54B applies where the protective conductor is not incorporated or bunched with cables, or for bare protective conductors in contact with cable covering.

54C applies where the protective conductor is a core in a cable or is bunched with cables.

This final value in ohms will now be added to Z_e, to give the value for the circuit of Z_s under full load conditions. (Z_e may have been measured, determined by calculation or ascertained by enquiry.) This value should then be compared with the requirements of Regulations 413-02-08 or 413-02-12 or 413-02-16 as appropriate to verify conformity.

413-02-08
413-02-12
413-02-16

Alternatively, as a rule of thumb, the measured value of earth fault loop impedance of each circuit at the most remote outlet should not exceed three-quarters of the relevant values in Table 41B or in Table 41D. If the measured value exceeds this comparative value, the reference test should be applied to identify the location of the high impedance.

Table 41B
Table 41D

Note: The normal method of test employed by a phase earth loop tester is to compare the unloaded loop circuit voltage with the circuit voltage when loaded with a low resistance, typically 10 ohms. This method of test can create an electric-shock hazard if the phase earth loop impedance is high and the test duration is not limited. In these circumstances the potential of the protective conductor could approach phase voltage for the duration of the test.

Section 15 — Functional Testing

15.1 Operation of residual current devices 713-12-01

In order to test the effectiveness of residual current devices after installation a series of tests must be applied to check that they are within specification and operate satisfactorily. This test sequence will be in addition to proving that the test button, if fitted to the rcd, is operational. The effectiveness of the test button should be checked after the test sequence.

Prior to these rcd tests it is essential, for safety reasons, that the earth loop impedance is tested to check the requirements have been met.

413-02-16

Instrument: Use an rcd tester for this test. Refer to Section 16.

Test method

The test is made on the load side of the rcd between the phase conductor of the circuit protected and the associated cpc. The load supplied should be disconnected during the test.

Note: RCD testers require a few milliamperes to operate the instrument, and this is normally obtained from the phase and neutral of the circuit under test. When testing a three-phase rcd protecting a three-wire circuit, its neutral is required to be connected to earth. This means that the test current will be increased by the instrument supply current and will cause some devices to operate during the 50 per cent test at a time when they should not operate. Under this circumstance it is necessary to check the operating parameters of the rcd with the manufacturers before failing the device.

413-02-15
413-02-19

Note: These tests can result in a potentially dangerous voltage on exposed and extraneous-conductive-parts when the earth fault loop impedance approaches the maximum acceptable limits. Precautions must therefore be taken to prevent contact of persons or livestock with such parts.

General purpose rcds to BS 4293

(i) with a fault current flowing equivalent to 50% of the rated tripping current of the rcd for a period of 2 s, the device should not open

(ii) with a fault current flowing equivalent to 100% of the rated tripping current of the rcd, the device should open in less than 200 ms.

Where the rcd incorporates an intentional time delay it should trip within the time range of 50% of the rated time delay plus 200 ms to 100% of the rated time delay plus 200 ms.

Because of the variability of the time delay it is not possible to specify a maximum test time. It is therefore imperative that the circuit protective conductor does not rise more than 50 V above earth potential ($Z_s I_n \leq 50$ V). It is suggested that in practice a 2 s maximum test time is sufficient.

413-02-16

General purpose rcds to BS EN 61008 or rcbos to BS EN 61009

(i) with a fault current flowing equivalent to 50% of the rated tripping current of the rcd for a period of 2 s, the device should not open

(ii) with a fault current flowing equivalent to 100% of the rated tripping current of the rcd, the device should open in less than 300 ms unless it is of "Type S" (or selective) which incorporates an intentional time delay. In this case it should trip within the time range of 130 ms and 500 ms.

RCD protected socket-outlets to BS 7288

(i) with a fault current flowing equivalent to 50% of the rated tripping current of the rcd for a period of 2 s, the device should not open

(ii) with a fault current flowing equivalent to 100% of the rated tripping current of the rcd, the device should open in less than 200 ms.

Additional requirement for supplementary protection

Where the rcd is used to provide supplementary protection against direct contact in accordance with Regulation 412-06-02, with a test current of 150 mA the device should open in less than 40 ms. The maximum test time must not be longer than 50 ms.

412-06-02

Integral test device

An integral test device is incorporated in each rcd. This device enables the mechanical parts of the rcd to be checked.

Tripping the rcd by means of the above electrical tests and the integral test device establishes the following:

(i) that the rcd is operating within the correct sensitivity

(ii) the integrity of the electrical and mechanical elements of the tripping device.

Operation of the integral test device does not provide a means of checking:

(i) the continuity of the earthing conductor or the associated circuit protective conductors, or

(ii) any earth electrode or other means of earthing, or

(iii) any other part of the associated installation earthing

(iv) the sensitivity of the device.

15.2 Functional checks
713-12-02

All equipment, including switchgear, controlgear, drives, controls and interlocks shall be checked and operated to confirm that they work and are properly mounted, adjusted and installed.

Section 16 — Test Instrument Requirements and Precautions

16.1 Instrument accuracy

A basic measurement accuracy of 5% is usually adequate for these test instruments. In the case of analogue instruments, a basic accuracy of 2% of full scale deflection will provide the required accuracy measurement over a useful proportion of the scale.

It should not be assumed that the accuracy of the reading taken in normal field use will be as good as the basic accuracy. The 'in-service reading accuracy' is always worse than the basic accuracy, and additional errors derive from two sources:

(i) *Instrument errors:*
basic instrument accuracy applies only in ideal conditions; the actual reading accuracy will also be affected by the operator's ability, battery condition, generator cranking speed, ambient temperature and orientation of the instrument, and may relate to loss of calibration. Regarding this latter point, instruments should be regularly recalibrated using standards traceable to National Standards, and should be checked after any mechanical or electrical mishandling

(ii) *Field errors:*
the instrument reading accuracy will also be affected by external influences as a result of working in the field environment. These influences may be in many forms, and the sources of such inaccuracies will be suggested in the appropriate sections.

To achieve satisfactory in-service performance, it is essential to be fully informed about the test equipment, how it is to be used, and the accuracy to be expected.

16.2 Low resistance ohmmeters

The instrument used for low resistance tests may be either a specialised low resistance ohmmeter, or the continuity range of an insulation and continuity tester. The test current may be ac or dc. It is recommended that it be derived from a source with no load voltage not less than 4 V and no greater than 24 V, and a short-circuit current not less than 200 mA.

713-02-01

The resolution at low scale values for instruments should be to 0.05 ohm or lower. Field effects contributing to in-service errors are contact resistance, test lead resistance, ac interference and thermocouple effects in mixed metal systems.

Whilst contact resistance cannot be eliminated with two terminal testers, and can introduce errors of 0.01 ohm or greater, the effects of lead resistance can be eliminated by measuring this prior to a test, and subtracting the resistance from the final value. Interference from an ac source cannot be eliminated, although it may be indicated by vibration of the pointer of an analogue instrument. Thermocouple effects can be eliminated by reversing the test probes and averaging the resistance readings taken in each direction.

16.3 Insulation resistance ohmmeters

The instrument used should be capable of developing the test voltage required across the load.

The test voltage required is:

(i)   250 V dc for extra-low voltage circuits

411-02-02 (i)

(ii)   500 V dc for all circuits rated up to and including 500 V, but excluding extra-low voltage circuits mentioned above

411-02-04

(iii)   1000 V dc for circuits rated above 500 V up to 1000 V.

The tester must be capable of supplying an output current of 1 mA at its nominal voltage so that it can develop its nominal test voltage when applied to the minimum acceptable insulation resistance.

The factors affecting in-service reading accuracy include 50 Hz currents induced into cables under test, and capacitance in the test object. These errors cannot be eliminated by test procedures. Capacitance may be as high as 5 μF, and any instrument should have an automatic discharge facility capable of safely discharging such a capacitance.

It may be helpful to note that instruments conforming to the German VDE 0413 Part 1 Standard will fulfil all the above instrument requirements.

16.4 Applied voltage testers

For the purposes of the applied voltage tests, the tester used should be capable of steadily increasing the test voltage with an accuracy of around 5 per cent. It should have means of indicating a failure of insulation, and should be capable of maintaining the test voltage continuously for at least 1 minute.

Care should be taken to prevent bare exposed metalwork becoming live unnecessarily, and the integrity of the earth connection of the test equipment should be verified.

Unless specifically required to overcome the capacitance in larger equipment or leakage that might be permissible with non-conducting floors and walls, the maximum output current of the tester should not exceed 5 mA. In general, the maximum test voltage required by the Wiring Regulations is 4000 V ac. If the maximum output of the test equipment can exceed 5 mA, it must be regarded as lethal and precautions should be taken to prevent danger to the person carrying out the test, and to anyone who might be near the apparatus or installation being tested.

16.5 Earth fault loop impedance testers

These instruments operate by circulating a current from the live conductor into the protective earth. This will raise the potential of the protective earth system.

To minimise electric shock hazard from the potential of the protective conductor, the test duration should be within safe limits. This means in general that the instrument should cut off the test current after a maximum of 40 ms or a time determined by the safety limits in IEC 479.

Instrument accuracy decreases as scale reading reduces. Aspects affecting in-service reading accuracy include transient variations of mains voltage during the test period, mains interference, test lead resistance and errors in impedance measurement as a result of the test method. To allow for the effect of transient voltages the test should be repeated at least once. The other effects cannot be eliminated by test procedures.

For circuits rated up to 50 A, a phase earth loop tester with a resolution of 0.01 ohms should be adequate.   413-02-07

16.6 Earth electrode resistance testers

The test should preferably be carried out using a 4 terminal tester (or a 3 terminal tester where a combined lead to the earth electrode would not have a significant resistance compared with the electrode resistance) so that the resistance of the test leads and temporary spike resistance can be eliminated from the test result.

Aspects affecting in-service reading accuracy include the effects of temporary spike resistance, interference currents and the layout of the test electrodes. Any instrument should carry some facility to check that the resistance to earth of the temporary potential and current spikes are within the operating limits of the instrument. It may be helpful to note that instruments complying with German VDE 0413 Part 7 standard will incorporate this facility. Care should be exercised to ensure that temporary spikes are positioned accurately.

The test instrument should be capable of applying the full range of test current to an in-service accuracy of 10 per cent. This in-service reading accuracy will include the effects of voltage variations around the nominal voltage of the tester.

To check rcd operation and to minimise danger during test the test current should be applied for no longer than 2 secs. Maximum test durations should be accurate to 5 per cent of required duration.

412-06

Section 17 — Forms of Completion and Periodic Inspection

17.1 Initial inspection and testing

Forms WR1 to WR4 are designed for use when inspecting and testing a new installation, or an alteration or addition to an existing installation. The Forms comprise the following:

WR1   Electrical Installation Completion Certificate from Appendix 6 of the Wiring Regulations

WR2   Form of Inspection

WR3   Form of Test

WR4   Form of Test Results and Installation Schedule

Guidance on completion is provided on the reverse of the Forms.

17.2 Periodic inspection

Form WR5 is a form for use when carrying out a routine periodic inspection and test of an existing installation. It is not for use when alterations or additions are made. Forms of Inspection (WR2), Test (WR3) and Test Results (WR4) are to accompany the Form of Periodic Inspection and Test (WR5).

17.3 Minor works

The complete set of forms for initial inspection and test may not be appropriate for minor works. When the addition to an electrical installation does not extend to the introduction of a new circuit the minor works form may be used. It is intended for such work as the addition of a socket-outlet or lighting point to an existing circuit.

Form WR6 is a typical form of completion and inspection for minor works.

COMPLETION CERTIFICATE AND PERIODIC INSPECTION REPORT FORMS

Introduction

(i)   The Completion Certificate required by Part 7 of BS 7671 shall be made out and signed by competent persons in respect of the design, construction, inspection and testing of the work.

(ii)   The Periodic Inspection Report required by Part 7 of BS 7671 shall be made out and signed by competent persons in respect of the inspection and testing of an installation.

(iii)   Competent persons will, as appropriate to their function under (i) and (ii) above, have a sound knowledge and experience relevant to the nature of the work undertaken and to the technical standards set down in this British Standard, be fully versed in the inspection and testing procedures contained in this Standard and employ adequate testing equipment.

(iv)   Completion Certificates will indicate the responsibility for design, construction, inspection and testing, whether in relation to new work or further work on an existing installation. For large or complex installations those responsible may provide an equivalent to this form, as identified in Regulation 741-01-01.

(v)   Periodic Inspection Reports will indicate the responsibility for the inspection and testing of an installation within the extent and limitations specified on the form.

(vi)   A schedule of test results as required by Part 7 shall be issued with the associated Completion Certificate or Periodic Inspection Report.

(vii)   When making out and signing a form on behalf of a company or other business entity, individuals shall state for whom they are acting.

(viii)   Additional forms may be required as clarification, if needed by non-technical persons, or in expansion, for larger or more complex installations.

ELECTRICAL INSTALLATION COMPLETION
CERTIFICATE (BS 7671 : 1992) (Notes 1 and 2)

Client's name/title: ....

DETAILS OF THE INSTALLATION

Tick boxes as appropriate

Installation Address: ....

....

New installation ☐	Extent of installation covered by this certificate
Addition to existing installation ☐	
Alteration to existing installation ☐	(Use continuation sheet if necessary)

PARTICULARS OF THE INSTALLATION

Type of Earthing: TN-C-S ☐   TN-S ☐   TT ☐   TN-C ☐   IT ☐

Details of Earth Electrode:

Type............... Location .... Method of Measurement .... Resistance .... Ω

Characteristics of the supply at the origin of the installation:

Nominal voltage V   Frequency ..... Hz   No of Phases .... Maximum demand (load) .... A per phase

	Measured	Calculated	Other
Maximum prospective fault current (note 7) kA			
External earth fault loop impedance, Ze Ω			

Overcurrent protective device at origin: Type: BS ....   Rating .... A

Main switch or circuit-breaker: Number of poles .... Type: BS ....   Rating .... A

(If a residual current device, rated residual operating current....................... mA)

Method of protection against indirect contact:

1. Earthed equipotential bonding and automatic disconnection of supply   ☐

2. Other (describe).. ☐

Main equipotential bonding conductors: Conductor material....................   csa mm2

COMMENTS ON EXISTING INSTALLATION, IN THE CASE OF AN ALTERATION OR ADDITION

DESIGN

I/We being the person(s) responsible (as indicated by my/our signatures below) for the design of the electrical installation, particulars of which are described on page 1 of this form CERTIFY that the said work for which I/we have been responsible is to the best of my/our knowledge and belief in accordance with BS 7671: 1992 - Requirements for Electrical Installations (16th Edition IEE Wiring Regulations), amended to....................(Note 3) except for the departures, if any, stated in this Certificate.

Details of departures (if any) from BS 7671: 1992 (120-02)

The extent of liability of the signatory is limited to the work described on page 1 of this form as the subject of this Certificate.

For the DESIGN of the installation

Name (IN BLOCK LETTERS): .... Position:...

Signature (Note 4): .... Date (Note 3):...

For and on behalf of:...

Address:...

.... Postcode......................

CONSTRUCTION

I/We being the person(s) responsible (as indicated by my/our signatures below) for the construction of the electrical installation, particulars of which are described on page 1 of this form CERTIFY that the said work for which I/we have been responsible is to the best of my/our knowledge and belief in accordance with BS 7671: 1992 - Requirements for Electrical Installations (16th Edition IEE Wiring Regulations), amended to....................(Note 3) except for the departures, if any, stated in this Certificate.

Details of departures (if any) from BS 7671: 1992 (120-02)

The extent of liability of the signatory is limited to the work described on page 1 of this form as the subject of this Certificate.

For the DESIGN of the installation

Name (IN BLOCK LETTERS): .... Position:...

Signature (Note 4): .... Date (Note 3):...

For and on behalf of:...

Address:...

.... Postcode...........................

INSPECTION AND TESTING

I/We being the person(s) responsible (as indicated by my/our signatures below) for the inspection and testing of the electrical installation, particulars of which are described on page 1 of this form CERTIFY that the said work for which I/we have been responsible is to the best of my/our knowledge and belief in accordance with BS 7671: 1992 - Requirements for Electrical Installations (16th Edition IEE Wiring Regulations), amended to....................(Note 3) except for the departures, if any, stated in this Certificate.

The extent of liability of the signatory is limited to the work described on page 1 of this form as the subject of this Certificate.

For the DESIGN of the installation

Name (IN BLOCK LETTERS): .... Position: ....

Signature (Note 4): .... Date (Note 3): ....

For and on behalf of: ....

Address: ....

.... Postcode ....

I/We RECOMMEND that the installation be further inspected and tested after an interval of not more than

NOTES ON COMPLETION CERTIFICATES

1. The Electrical Installation Completion Certificate is to be used only for the initial certification of a new installation or for an alteration or addition to an existing installation carried out in accordance with BS 7671: 1992. **It is not to be used for a Periodic Inspection** for which a Periodic Inspection Report form should be used. The **original** Certificate is to be given to the person ordering the work (Regulation 741-01-01). A **duplicate** should be retained by the contractor.

2. This Certificate is only valid if accompanied by the Schedule(s) of Test Results (see Note 6).

3. Dates to be inserted.

4. The signatures appended are those of the persons authorised by the companies executing the work of design, construction and inspection and testing respectively. A signatory authorised to certify more than one category of work should sign in each of the appropriate places.

5. The time interval recommended before the next periodic inspection must be inserted (see IEE Guidance Note No. 3 for guidance).

6. The page numbers for each of the Schedules of Test Results should be indicated, together with the total number of sheets involved.

Page 3 of   (note 6)

INSPECTION SCHEDULE

Every Installation shall during erection and/or on completion and before being put into service be inspected and tested to verify, so far as is reasonably practicable, that the requirements of the Regulations are being met.

Items Inspected

☐ 1. Connection of conductors

☐ 2. Identification of conductors

☐ 3. Routeing of cables in safe zones or protected against mechanical damage

☐ 4. Selection of conductors for current and voltage drop

☐ 5. Connection of single-pole devices for protection or switching in phase conductors only

☐ 6. Correct connection of socket-outlets and lampholders

☐ 7. Presence of fire barriers and protection against thermal effects

8. Method of protection against electric shock

   a) protection against both direct and indirect contact

☐   SELV

☐   Limitation of discharge of energy

   b) protection against direct contact

☐   Insulation of live parts

☐   Barrier or enclosure

☐   Obstacles

☐   Placing out of reach

☐   PELV

   c) Protection against indirect contact

   (i) Earthed equipotential bonding and automatic disconnection of supply

☐   Presence of earthing conductors

☐   Presence of protective conductors

☐   Presence of main equipotential bonding conductors

☐   Presence of supplementary equipotential bonding conductors

   (ii) Use of Class II equipment or equivalent Insulation

☐   (iii) Non-conducting location

   Absence of protective conductors

   (iv) Earth-free local equipotential bonding

☐   Presence of earth-free equipotential bonding conductors

   (v) Electrical separation (413-06)

9. Prevention of mutual detrimental influence

☐   Proximity of non-electrical services and influences

☐   Separation of Category 1 and Category 2 cables or Category 1 insulation used

☐   Separation of Category 3 cables

☐ 10. Presence of appropriate devices for isolating and switching correctly located

☐ 11. Presence of undervoltage protective devices where appropriate

12. Choice and setting of protective and monitoring devices (for protection against indirect contact and/or overcurrent)

☐   Residual current devices

☐   Overcurrent devices

☐ 13. Labelling of protective devices, switches and terminals

☐ 14. Selection of equipment and protective measures appropriate to external influences

☐ 15. Adequacy of access to switchgear and equipment

☐ 16. Presence of danger notices and other warnings

☐ 17. Presence of diagrams instructions and necessary information

☐ 18. Erection methods

☐ 19. Requirements of special locations

Tick to indicate satisfaction with inspection
Delete if item not applicable

Inspected by ....

Date...

NOTES ON INSPECTIONS - FORM WR2

8a   SELV An extra-low voltage system which is electrically separate from earth and from other systems. The particular requirements of the Regulations must be checked (see Regulation 411-02)

8b   Method of protection against direct contact — will include measurement of distances where appropriate

Obstacles — only adopted in special circumstances (see Regulation 412-04)

Placing out of reach — only adopted in special circumstances (see Regulation 412-05)

8c   Use of Class II equipment — infrequently adopted and only when the installation is to be supervised (see Regulation 413-03)

Non-conducting locations — not applicable in domestic premises and requiring special precautions (see Regulation 413-04)

Earth-free local equipotential bonding — not applicable in domestic premises, only used in special circumstances (see Regulation 413-05)

Electrical separation (see Regulation 413-06)

TEST SCHEDULE

SCHEDULE OF ITEMS TO BE TESTED

1. ☐ Continuity of protective conductors

2. ☐ Continuity of ring final circuit conductors

3. ☐ Insulation resistance between live conductors and earth

4. ☐ Site applied insulation

5. ☐ Protection by separation of circuits

6. ☐ Protection against direct contact, by barrier or enclosure provided during erection

7. ☐ Insulation of non-conducting floors and walls

8. ☐ Polarity

9. ☐ Earth electrode resistance

10. ☐ Earth fault loop impedance

11. ☐ Operation of residual current-operated devices

12. ☐ Functional testing of assemblies

Tick items meeting test requirement. Delete tests not applicable

DEPARTURES FROM REGULATIONS (note should be made of compliance with relevant British Standards where appropriate)

....

....

....

ITEMS NOT TESTED

....

....

....

FORM OF INSPECTION No../2

SCHEDULES OF TEST RESULTS Form No../4

I/We have carried out the inspection and test of the electrical installation and certify that the inspection and tests indicate to the best of my/our knowledge and belief that the installation is in accordance with the Regulations for Electrical Installations 16th Edition as amended on .... except for departures listed above.

Name (in Block Letters) ....   Position ....

Signature ....

For and on behalf of ....

Address ....

       ....

       ....

FORM WR4   Form No xxx/4

INSTALLATION SCHEDULE (including Test Results)

Contractor...

Test Date...   Address/Location of dis. board   1 Type of Supply TN-S/TN-C-S/TT   Instruments:
                                                         ....   2 Z_e at origin..............................ohms   rcd tester.....................................
Signature...   ....   3 PSSC ....kA   continuity....................................
                                                         ....   insulation...................................
                                                                                                                                                          others..

Equipment vulnerable to testing..

Description of work completed																					
		Overcurrent Device			Circuit Conductors		Test Results											
Circuit	no of points	type	Rating	Short-circuit capacity	live	CPC	Continuity		Insulation Resistance				Earth Loop Imped-ance	Functional Testing			Polarity	Remarks
							$R_1 + R_2$	R_2	Phase/ Phase	Phase/ Neutral	Phase/ Earth	Neutral/ Earth		RCD		Other		
			A	kA	mm^2	mm^2	Ω	Ω	$M\Omega$	$M\Omega$	$M\Omega$	$M\Omega$	$Z_t\ \Omega$	$I_{\Delta n}$	time ms			
4	5	6	7	8	9	10	11	12	13	14	15	16	17	18	19	20	21	22
1																		
2																		
3																		
4																		
5																		
6																		
7																		
8																		
9																		
10																		
11																		
12																		

22 Main bonding check: Gas..... Water..... Other..... size............mm^2 23 Main Earth. size............ mm^2 24 Earth Electrode Resistance......................Ω

Deviations from Wiring Regulations and special notes:

NOTES ON TEST AND INSTALLATION SCHEDULE

Tests shall be carried out in the sequence below and results recorded on the installation schedule form WR4.

1   Continuity of protective conductors

Every protective conductor including bonding conductors shall be tested to verify it is sound and correctly connected. (See note 2)

2   Continuity of final circuit conductors

A test shall be made to verify the continuity of each conductor including the protective conductor of every ring circuit.

The sum of the resistance of the phase conductor and the protective conductor $(R_1 + R_2)$ is to be inserted in column 13 of the installation schedule. After temperature correction this may be used, by the addition of Z_e, to determine Z_s.

Where the alternative method of Regulation 413-02-12 is used for shock protection the resistance of the circuit protective conductor R_2 is measured and recorded in column 12.

3   Insulation resistance

Electronic devices shall where necessary be disconnected from the installation so that they are not damaged by the testing. They must be recorded on the installation schedule and test schedule.

Where the devices have exposed-conductive-parts the insulation resistance between the exposed-conductive-parts and phase and neutral conductor connected together shall be measured. It must comply with the appropriate British Standard or if there is no standard be not less than 0.5 megohm.

The insulation resistance between phase/neutral, phase/earth, and neutral/earth shall be measured and the values recorded in columns 14,15 and 16. The minimum insulation resistance required for the main switchboard, and each distribution circuit tested separately with all final circuits connected, but current using equipment disconnected, is as Table 71A in the Wiring Regulations.

4   Site applied insulation

Where insulation is applied on site it shall be capable of withstanding a test voltage equivalent to that required by the British Standard for similar type tested equipment. The insulated enclosure must provide a degree of protection not less than IP2X.

5   Protection by separation of circuits

Where protection is provided by SELV see Regulations 411-02 and 471-02, where provided by electrical separation see Regulation 413-06 and 471-12 (eg isolating transformer).

6   Protection against direct contact by barrier or enclosure provided during erection — see Regulation 412-03

7   Insulation of non-conducting floors and walls

See Regulations 413-04 and 471-10.

8   Polarity

It shall be verified that:

(a) every fuse and single-pole control and protective device is connected in the phase conductor only
(b) centre contact bayonet and Edison screw lampholders have outer contact connected to the neutral conductor
(c) wiring is correctly connected to socket-outlets and similar accessories.

Compliance is to be indicated in column 20 of WR4.

9   Earth electrode resistance

The earth electrode resistance of TT installations must be measured, and normally an rcd is required.

For reliability in service the resistance of any earth electrode should be below 200 Ω. Record the value in column 24 of Form WR4.

All the preceding tests should be carried out before the installation is energised.

10 Polarity

Following energising of the installation, polarity must be checked before further testing.

11 Earth fault loop impedance Z_t

This will be measured at the furthest point on each circuit and recorded in column 17 of the schedule. The maximum value of Z_t is provided by the installation designer or may be assumed to approximate to Z_s given by Tables 41B1, 41B2 and 41D multiplied by 0.96/120 (for an ambient of 10°C and cables to Table 54C). When Z_s is determined by limiting circuit length, the length shall be recorded in column 11 of Form WR4.

12 Functional testing

The operation of rcds and rcbos shall be tested by simulating a fault condition, independent of any test facility in the device. At rated tripping current rcds must operate within 200 msec. Record operating current and time in column 18 and 19. Effectiveness of the test button must be confirmed.

All switchgear and controlgear, drives, interlocks etc shall be operated to ensure they are properly mounted, adjusted, are safe and work. Satisfactory operation is indicated by a tick in coloumn 20.

Prospective Fault Current. This may be measured or determined by calculation, or ascertained by enquiry. The value is inserted in the form headed 'Particulars of the Installation'.

110

PERIODIC INSPECTION REPORT FOR AN ELECTRICAL INSTALLATION (BS 7671 : 1992) (Note 1)

DETAILS OF THE CLIENT

Client: ....

Address: ....

Purpose for which this Report is required: ....(Note 3)

DETAILS OF THE INSTALLATION

Occupier: ....

Address: ....

....

| Domestic | Commercial | Industrial |
| Other | | |

Description of Premises:

Estimated age of the Electrical Installation: .... years

Evidence of Alterations or Additions: | Yes | No | Not Apparent |

If "Yes", Estimate Age: .... years

Date of last inspection: ....   Records available | Yes | No |

Records held by:

Type of Earthing: TN-C-S ☐   TN-S ☐   TT ☐   TN-C ☐   IT ☐

Details of Earth Electrode:

Type ....Location...................... Method of Measurement .... Resistance.............Ω

Characteristics of the supply at the origin of the installation:

Nominal voltage V   Frequency ..... Hz   No of Phases .... Maximum demand (load) ..... A per phase

	Measured	Calculated	Other
Maximum prospective fault current (note 7) kA			
External earth fault loop impedance, Ze Ω			

Overcurrent protective device at origin: Type: BS ....   Rating .... A

Main switch or circuit-breaker: Number of poles .... Type: BS ....   Rating .... A

(If a residual current device, rated residual operating current .... mA)

Method of protection against indirect contact:

1. Earthed equipotential bonding and automatic disconnection of supply   ☐

2. Other (describe) ....   ☐

Main equipotential bonding conductors: Conductor material .... csa mm2

EXTENT AND LIMITATIONS OF THE INSPECTION

Extent of Electrical Installation Covered by this Report (Note 5):..

....

Limitations: ....

....

RECOMMENDATIONS (Note 9)

Referring to the "Schedule(s) of Inspection and Test Results", and subject to the limitations specified above — Recommendations as detailed below

☐ No remedial work required, or

The following items:..

....

....

....

One of the following numbers shall be placed alongside each of the items detailed above (Note 6).

| 1 | requires urgent attention | 2 | requires improvements | 3 | requires further investigation |

| 4 | does not comply with BS 7671: 1992 (as amended) (This does not necessarily imply that the electrical installation is unsafe). |

SUMMARY OF THE INSPECTION

Date(s) of the inspection: ....

General condition of the installation (Note 7): ....

....

....

....

....Overall assessment: Satisfactory/Unsatisfactory

(Note 2)

SCHEDULE OF THE INSPECTION: See Sheet(s) ....attached

SCHEDULE OF TEST: See Sheet(s) ....attached

SCHEDULE OF THE INSPECTION AND TEST RESULTS: See Sheet(s) ....attached

NEXT INSPECTION

I/We recommend that the installation should be re-inspected after an interval of not more than..... months/years (Note 8)

DECLARATION

To the best of our knowledge and belief I/We confirm that the details recorded above and in the attached Schedule(s) of Inspection and Test Results and the Recommendations are an accurate assessment, within the limits specified, of the condition of the electrical installation as above.

INSPECTED BY: ....   REVIEWED BY:

Signature:..   Signature: ....

Name (IN BLOCK LETTERS): ....   Name (IN BLOCK LETTERS):........................

Date of signing:..   Date of signing ....

For and on behalf of: ....

Address: ....

NOTES:

1. This Periodic Inspection Report form shall be used only for reporting on the condition of an existing installation.

2. The Report, shall include schedules of both the inspection and the test results. Additional sheets of test results may be necessary for other than a simple installation. The page number of each sheet shall be indicated, together with the total number of sheets involved.

3. The intended purpose of the Periodic Inspection Report shall be identified, together with the recipient's details in the appropriate boxes.

4. The maximum prospective fault current recorded should be the greater of either the short-circuit current (between the live conductors) or the earth fault current (between phase conductor and an exposed-conductive-part).

5. The 'Extent and Limitations' box shall fully identify the elements of the installation that are covered by the report and those that are not; this aspect having been agreed with the client and other interested parties before the inspection and testing is carried out.

6. The recommendation(s), if any, shall be categorised using the numbered coding 1-4 as appropriate against each recommendation.

7. The 'Summary of the Inspection' box shall clearly identify the condition of the installation in terms of safety.

8. Where the periodic inspection and testing has resulted in a satisfactory overall assessment, the time interval for the next periodic inspection and testing shall be given. The IEE Guidance Note No. 3 provides guidance on the maximum interval between inspections for various types of buildings. If the inspection and test reveals that parts of the installation require urgent attention, it would be appropriate to set an earlier re-inspection date having due regard to the degree of urgency and extent of the necessary remedial work.

9. If the space available on the model form for information on recommendations is insufficient, additional pages shall be provided as necessary.

Page 3 of   (note 2)

MINOR WORKS

To be used only for the completion and inspection certification of electrical installation work which does not include a new circuit.

PART 1 CERTIFICATE

I/We being responsible for the minor electrical works carried out at:

and described below, certify that the said works do not impair the safety of the existing installation and the said works have been inspected and tested, in accordance with the BS7671: 1992, IEE Wiring Regulations Sixteenth Edition and that, to the best of my/our knowledge and belief, at the time of my/our inspection complied with these Regulations except as detailed at PART 4 below.

(i)   Name (In Block Letters):   (ii) Position:

(iii)   For and on behalf of:

        Address:

        Signature: Date:

PART 2 DESCRIPTION OF MINOR WORKS

(i)   Date minor works carried out:....................................

(ii)   Contract reference:....................................

(iii)   Description of the minor works:

PART 3 TESTING

Essential Tests:

(i) Earth Continuity	
(ii) Insulation Resistance Phase/Earth and Neutral/Earth MΩ	
(iii) Phase/Earth Loop Impedance Ω	
(iv) Polarity	
(v) RCD Operation	

PART 4 EXISTING INSTALLATION

(i)   Method of Protection against indirect contact:

        a) Earthed equipotential bonding and automatic disconnection of supply:

        b) Other: describe:

(ii)   Earthing adequate/not adequate:

(iii)   Main Equipotential bonding: adequate/not adequate

(iv)   Comments on Existing Installation:

MINOR WORKS

Scope

(i)   The minor works form is only to be used for additions to an electrical installation that do not extend to the introduction of a new circuit eg the addition of a socket-outlet or a lighting point to an existing circuit.

Part 1 Declaration

(i)   The Form required by part 7 shall be made out and signed by competent persons in respect of the design, construction, inspection and testing of the work

(ii)   The competent person will have a sound knowledge and experience relevant to the nature of the installation undertaken and to the technical standards set down in the Wiring Regulations, be fully versed in the inspection and testing procedures contained in the Regulations and employ adequate testing equipment

(iii)   When making out and signing a form on behalf of a company or other business entity, individuals shall state for whom they are acting.

Part 2 Description of the Works

(i)   The minor works must be so described that the work that is the subject of the certification can be readily identified.

Part 3 Testing

The relevant provisions of Part 7 (Inspecting and Testing) of the 16th Edition IEE Wiring Regulations must be applied in full to all minor works. For example where a socket-outlet is added to an existing circuit it is necessary to:

(i)   establish that the earthing contact on the socket-outlet is connected to the main earthing terminal via a low impedance

(ii)   measure the insulation resistance of the circuit that has been added to, and establish that it complies with Table 71A in the 16th Edition, IEE Wiring Regulations

(iii)   Measure the earth fault impedance to establish that the maximum permitted disconnected time is not exceeded

(iv)   check that the polarity of eg the socket-outlet, is correct

(v)   if the work is protected by an rcd, the effectiveness of the rcd must be verified, and recorded.

Part 4 Existing Installation

(i)   The method of protection against indirect contact shock hazard must be clearly identified

(ii)   If the existing installation lacks either an effective means of earthing or adequate main equipotential bonding conductors, this must be clearly stated

(iii)   Recorded departures from the 16th Edition IEE Wiring Regulations may constitute non-compliance with the Electricity Supply Regulations 1988 as amended or the Electricity at Work Regulations 1989. It is important that the client is advised immediately in writing.

Index

NOTES

NOTES

NOTES

NOTES

NOTES

NOTES

NOTES

NOTES

NOTES

NOTES

NOTES

NOTES